# THE FLAVOR MATRIX

## JAMES BRISCIONE

WITH BROOKE PARKHURST

HOUGHTON MIFFLIN HARCOURT

BOSTON  NEW YORK  2018

# THE ART
## AND SCIENCE
### OF
## PAIRING COMMON INGREDIENTS TO CREATE EXTRAORDINARY DISHES

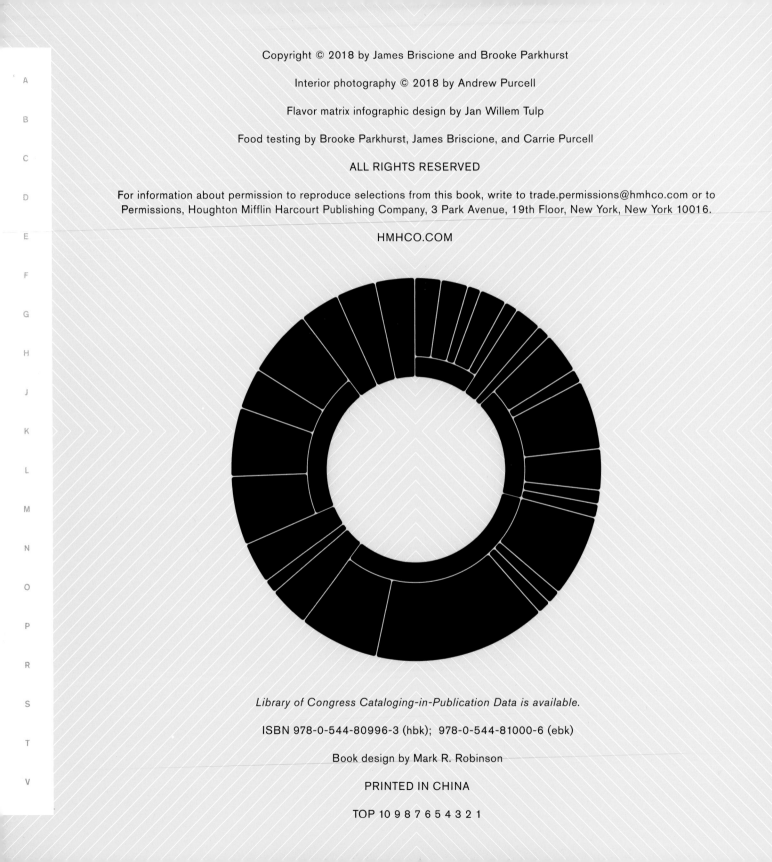

*Library of Congress Cataloging-in-Publication Data is available.*

ISBN 978-0-544-80996-3 (hbk);  978-0-544-81000-6 (ebk)

Book design by Mark R. Robinson

PRINTED IN CHINA

TOP 10 9 8 7 6 5 4 3 2 1

A
B
C
D
E
F
G
H
J
K
L
M
N
O
P
R
S
T
V

# Dedication

To Brooke,

Most people tell us they can't imagine working on a project like this with a spouse.

I can't imagine achieving this with anyone else. Thank you for your support, motivation

(both gentle and less so), and, most of all, your love.

—James

For "Jamie Chef," always.

—Brooke

# INTRODUCTION

# The Art and Science of Flavor Pairing

The idea for this book came to me while I was cooking with a computer.

The year was 2012, and my kitchen companion was Watson, IBM's famous supercomputer. If you're like most people, you first met Watson when it competed—and dominated—on the TV show *Jeopardy!* You might not have known that the computer's talents extend beyond trivia. I certainly didn't when I first met Watson—but I was in for a surprise.

When IBM had approached the Institute of Culinary Education, where I work, to collaborate on a project that would mix computing and human creativity in a way that had never been attempted before, I was honored to be asked to participate, but I was also very skeptical. How could a trivia-playing computer teach me anything about cooking? After all, I had devoted my life to being at the cutting edge of the culinary world. Besides, I thought, cooking is a craft—an art, something that engages the senses and requires a human touch. What could a computer know about the way a ripe tomato yields to the touch, or what a steak sounds like as it sizzles to perfect doneness?

It wasn't long before I realized that Watson was capable of much greater things than simply answering questions. Since proving its revolutionary computing abilities on *Jeopardy!*, Watson had turned its prodigious talents to managing care for cancer patients, understanding financial markets, helping law enforcement solve cold cases, even providing music recommendations. Now Watson was going to attempt to help seasoned chefs become more creative in the kitchen. *Jeopardy!* had proven that Watson could answer any question posed to it, but what about the questions we don't even know to ask?

The collaboration would work like this: Watson would create a list of ingredients that I or another of the chef-instructors at ICE would then use to create a dish with ingredient combinations that had rarely—if ever—been seen before. The foods Watson would select for us often seemed random at first glance, but would be anything but; rather, through a sweeping analysis of data pulled from academic articles, cookbooks, and other sources, Watson would find combinations of ingredients that it predicted would taste good when combined—but that did not commonly appear together in recipes.

I enjoyed cooking with Watson. We produced some interesting dishes: roast duck with tomato, sage, olives, and cherries; lobster with pork, saffron, basil, carrots, and balsamic vinegar; chicken with strawberries, mushrooms, and apple. These were combinations that I never would have come up with on my own, and our work together inspired me to reconsider what creativity in the kitchen actually looks like. But I knew I had only scratched the surface.

Watson didn't view ingredients the same way that I did as a chef. When I look at a tomato, I immediately think about all the ingredients I have tasted with tomato in the past, too often focusing on those that are familiar and comfortable like basil, cheese, or olives. Watson, on the other hand, thought about combining ingredients based only on their inherent flavors, with no notions of which foods conventionally go together. The computer seemed to be able to see the invisible filaments that bound different ingredients together.

For the moment, the ability to see these connections was the exclusive domain of scientists and supercomputers. But I wanted to learn to see them for myself, and to help others do the same. None of my students at ICE, let alone my family or friends, had access to a supercomputer that could help them think more creatively about flavor pairing as Watson had helped me do. How could I share this amazing experience with them?

Luckily, there is a vast body of scientific data out there about flavor, just waiting to be explored. But this information isn't part of the language of chefs—which is a shame, because the art and (yes) science of food pairing is in some ways the final frontier of modern cooking.

The world of food has been revolutionized by science in the past few decades. Since the concept of molecular gastronomy was invented in the late 1980s, sophisticated new ingredients, devices, and techniques—from sodium citrate to sous vide machines to spherification—have moved from elite restaurants to fast-casual service counters and home kitchens. Many people now cook with an understanding of how their efforts are transforming ingredients at a molecular level. The impact that science has had on the world of food cannot be overstated.

Except in one regard: In the midst of this epic disruption, the way we think about combining ingredients themselves has remained largely unchanged.

The old model of combining ingredients, of pairing their flavor profiles, effectively shackles chefs to ingredients, flavors, and combinations they already know. Many chefs and ambitious home cooks have the ability to flip through a mental Rolodex of ingredients, instantly recalling which ingredients and flavors work well together. This ability, the result of an intimate knowledge of food built up over years of cooking and eating, is called taste memory. But even the most extensive taste memory has its limitations. It's great for cataloging what combinations have been enjoyed in the past, but when it comes to deciding what could go together, taste memory is limited by the cook's personal experience.

You could spend a lifetime familiarizing yourself with the tens of thousands of ingredients available around the world. The challenge is especially daunting given how many there are to choose from right in our own neighborhoods. Between 1980 and 2014, for example, the number of items in the average American grocery store tripled, from 15,000 to around 44,000, according to the Food Marketing Institute. Farmers' markets, meanwhile, which may be stocked with unique heirloom fruits and vegetables that many of us have never heard of before, are on the rise. USDA statistics show that in 1994 there were 1,755 farmers' markets across the United States. Twenty years later, the country had over 8,250 active markets.

All of this choice poses an acute challenge for the modern cook. Whether we're overwhelmed by the sheer volume of our options, or simply confronting a mishmash of strange ingredients in an otherwise empty fridge, how are we to decide which of these ingredients we can put together in a way that will be interesting and also taste good?

There are other books that list ingredients commonly used in the same dishes, but few have ventured beyond convention to explore the untapped potential of how ingredients could be used together. This kind of mapping and exploration is exactly what I aim to do in the pages ahead. And rather than rely on taste memory, I'm going to rely on chemistry—specifically, on an analysis of ingredients' shared molecular structures, one of the most exciting areas of food science today.

Over the past thirty years, researchers have changed the way we think about flavor. Food scientists have taught chefs that each ingredient has a complex network of chemical structures called volatile compounds that give each food its own unique flavor. These compounds are responsible for 80 percent of what we perceive when we take a bite or a sip. A simple ingredient like lettuce has about twenty such compounds. Coffee has nearly 1,000.

Think of each of these volatile compounds as a single type of pixel in a photo—all the pink pixels, say. Alone, these pixels are just isolated colored squares. But when they're put together with hundreds of other colored squares, an image emerges. Create the image without the pink squares and your picture will be incomplete, no matter whether it shows a sunset, or a forest, or the inky depths of outer space.

In the same way, a single volatile compound does not give us the whole picture of the product in which it's found. For example, mesifurane—a volatile compound prominent in strawberries—does not taste like fruit at all. It has the aroma of baked bread, butter, and toasted almond. Yet if mesifurane were removed from a strawberry, the fruit's taste would change dramatically. Almond croissants with strawberry jam wouldn't be the same, either.

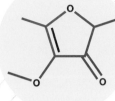

**MESIFURANE**

Volatile compounds such as mesifurane are the essential building blocks of flavor. So you may be surprised to learn that they have little to do with taste.

The words "taste" and "flavor" are often used interchangeably. But in fact they refer to fundamentally different things. Understanding the distinction is a crucial first step to mastering the art and science of pairing ingredients.

There are five—or six, depending on whom you ask—basic tastes: salty, sweet, sour, bitter, umami, and fat. We perceive each of these tastes via chemical reactions that take place on the tongue when we consume food. Chemical receptors, commonly known as taste buds, are distributed over the tongue as well as the soft palate, upper esophagus, cheeks, and even the epiglottis (the piece of cartilage that covers your windpipe when you swallow). About half of our taste buds contain sensors for each taste (while others are dedicated to specific tastes) and will fire signals back to the brain whenever it comes in contact with any of the five—or six—tastes. Contrary to popular wisdom, while certain parts of the tongue may be more sensitive to specific tastes, any taste can be perceived on any part of the tongue.

But these tastes are only 20 percent of what we perceive as flavor. That's right: Just one-fifth of what we experience when taking a bite of food comes from taste buds. The other 80 percent is reported by the nose. And while the taste component of flavor is created by familiar molecular elements in our food, such as acids, sugars, fats, and proteins, the aroma component comes from an entirely different class of molecules: those volatile compounds I mentioned earlier.

Every bite of food contains hundreds if not thousands of volatile compounds, which I will refer to as aromatic compounds. And as that name suggests, it is the smell of these compounds that dictates flavor.

When it comes to enjoying food, our sense of taste would be nothing without our sense of smell. Test that for yourself: Next time you sip your coffee, pinch your nose closed and see if it doesn't just taste like warm, slightly bitter water. Without your nose to sense the hundreds of aromatic compounds, coffee—or any substance—becomes nearly flavorless.

All aromas contribute to flavor in the same way. Before a single bite enters your mouth, aromatic compounds make their way to receptor cells in your nose and throat. When specific compounds reach their unique receptor cells (imagine square-, round-, triangular-, and star-shaped pegs looking to fit into their corresponding holes), they bond with these cells and fire a signal to the brain. There it is translated into the compounds' distinctive aromas. This takes place almost instantaneously, and can happen at a great distance. Think about how much the aromas of burgers on a grill tell you about the food's flavor as the scent wafts through the summer air, even if you're sitting on the other side of a baseball field from where the food is being cooked.

Once the food enters your mouth, chewing it releases yet more aromatic compounds; over a thousand individual compounds may be found in a single bite. These signals, too, register in your nose and throat. Your brain combines this data with information from taste buds about what you are tasting.

Signals from the vibrations in your jaw and eardrums report data on a food's texture to your brain, rounding out your perception of flavor. In controlled experiments, researchers have shown that texture specifically affects perception of salt and sweetness, but can also affect perception of flavor more generally.

# Under the Microscope:
# Volatile Versus Aromatic Compounds

Broadly speaking, an "aromatic" compound is any chemical that we can smell. For chemists, however, an aromatic compound has an even more specific definition. When they were first discovered, aromatic compounds such as benzene were noted because they had properties different from other hydrocarbons (see page 256). Specifically, they had a different kind of odor than other compounds, and were named aromatic because of their smell. A volatile compound, on the other hand, is any chemical that moves into the air, whether it has a smell or not. Water is an example of volatile compound with no odor, while benzene is both aromatic and volatile.

**BENZENE**

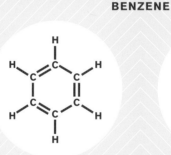

FOUND IN GINGER, MUSSELS, EGGPLANT, AND MUCH MORE

Both volatile and aromatic compounds play an important role in creating flavor in food. For our purposes, it is of little consequence whether a compound is technically volatile or aromatic. What matters is that it has a smell—and therefore a flavor. For the sake of simplicity, in this book I refer to all such compounds as aromatic.

All three factors—taste, aroma, and texture—help form a complete picture of flavor. But this one sense, aroma, is far more influential than the others.

Despite the crucial role that aroma plays in creating food's flavor, we tend to save descriptions of it for talking about wine. A sommelier can rattle off more than a hundred different descriptors for the aromas found in wine. Although the same aromas, plus many more, are found in food, and although they translate almost directly into flavors, our vocabulary for flavors in food rarely ventures beyond the five or six basic *tastes*.

If we put the same thoughtfulness into categorizing the flavors we find in food that we use for the aromas we find in wine, our culinary experiences would be heightened immeasurably. And if we gave the same care and attention to pairing ingredients with each other that we give to pairing wine with meals, the resulting dishes could be truly revolutionary—to say nothing of the resulting wine pairings!

Of course, categorizing flavors is far more complicated than categorizing tastes. There are just so many more of the former than the latter. The tastes we experience are the product of a limited number of chemical receptors in our mouths, while the flavors we find in food are comprised of thousands of aromatic compounds. Just as you could spend a lifetime building your taste memory, you could spend a lifetime (or several) studying all the different flavors and aromas out there. Indeed, there are brilliant scientists who have devoted their entire careers to exactly that.

I am not one of these scientists. I'm just a chef who wants to understand how flavor works so I can be a better cook. But I wasn't going to let a lack of academic credentials stop me.

My desire to master the science of flavor pairing led me down a rabbit hole. I immersed myself in articles from academic journals, entries in online chemistry databases, and the theories of the most brilliant minds in the world of molecular gastronomy. Follow me while I explain the basics of what I found, and how this raw information became the book you hold in your hands.

After giving the matter some thought, I realized that if I organized the vast number of aromatic compounds into manageable categories, and measured these categories' presence in a range of different ingredients, I could create standardized flavor profiles of each ingredient that I could then compare to each other. Essentially I would be distilling the flavor of a given ingredient—often the result of hundreds of different compounds—into a visual format that could be easily deciphered and compared to other ingredients. With this information I could begin creating flavor pairings based on science rather than opinion.

As the first step in this process, I first had to broadly categorize the different aromatic compounds that are most commonly found in food. (You can see these different categories in the back of the book, starting on page 256.) This is harder than it sounds; each group of aromatic compounds I identified actually contains thousands of different compounds with a wide range of scents. Assigning a specific aroma profile to any group of compounds with complete accuracy is impossible. Some sort of editorial judgment is needed, but that seems right and proper for a book written by a human chef instead of by a scientist—or a computer, for that matter.

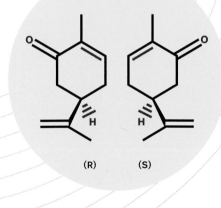

**R & S ISOMERS OF CARVONE**

An additional complication for this phase of my research was the fact that compounds with very similar shapes can have wildly varying aromas. For example, the compound carvone exists in two different configurations—the R isomer and S isomer. These are exact mirror images of one another (see opposite). The R isomer of carvone has an intense spearmint aroma; it is the main flavoring in Wrigley's famous chewing gum. However, the S isomer has the aroma of caraway seeds. Their structures are mirrored images, yet the compounds have unique aromas.

Luckily, basic chemistry offers a solution to this problem of categorization. Although the shape of a compound won't necessarily tell us what it smells like, the properties of specific *groups* of compounds can give us some clues about their aromas. This is because aromatic compounds can have distinguishing elements known as functional groups.

Functional groups help to classify aromatic compounds. The basic structure of any compound is a chain of carbon and hydrogen atoms. Many aromatic compounds are made up of more than just carbon and hydrogen, however; these additional atoms or groups of atoms are the functional groups. (An in-depth look at functional groups in aromatic compounds can be found beginning on page 256.) When different functional groups are detected by the receptor cells in the nose and throat, those sensors signal specific scents to the brain. For example, thiols (sulfur-containing functional groups) are stinky; at high concentrations, they smell of rotten eggs.

After reaching this realization about the basic aromatic characteristics of functional groups, I set about classifying the smells of food that are determined by these aromatic compounds. I divided these smells into primary aroma categories. When applicable, I also distinguished any notable subcategories of these aromas. (The list of aroma categories and subcategories I used is in the back of the book, on page 260.)

In addition to categorizing the main groups of primary aromas and their underlying aromatic compounds, I also pinpointed the different levels of aromatic compounds in roughly 120 of the most common ingredients in the modern kitchen: from groups of ingredients such as alliums (which include onions, garlic, chives, and ramps) to specific, unique ingredients such as truffle and vanilla. Research journals contain detailed analyses of the hundreds of different aromatic compounds found in each of the ingredients, but I found the bulk of this data using the Volatile Compounds in Food (VCF) database, managed by Triskelion B.V., an independent research facility located in the Netherlands. The VCF database is the most comprehensive listing of aromatic compounds in foods available in the world today. I supplemented the information I found there with scientific papers on aroma research and volatile compounds.

Once I completed this process, I was left with spreadsheet upon spreadsheet cataloging the significant aromatic compounds in each of the ingredients that I wanted to study. Finally, I could compare them to each other.

That comparison boiled down to a single number: the percentage of aromatic compounds that two different ingredients (or groups of ingredients) have in common. According to a relatively recent culinary innovation called flavor pairing theory, this simple metric can allow us to judge the relative compatibility of ostensibly dissimilar ingredients.

Flavor pairing theory—which, like all great theories, is elegantly simple—tells us that if two ingredients share significant quantities or concentrations of aromatic compounds, they will likely taste good together when combined in a dish. Sports fans can think of this as the analytics of flavor (and flavor pairing theorists like me are your Billy Beanes).

This powerful theory originated in 1999 with the pioneering chef Heston Blumenthal and his research team at The Fat Duck, a restaurant in Bray, England, with its own dedicated research kitchen (which is more like a lab than most kitchens you've seen). Blumenthal and his colleagues stumbled upon their theory while experimenting with salty and sweet foods. They combined caviar and white chocolate, discovering to their surprise that the pairing tastes delicious.

# Ingredient Groups

You may come across some unfamiliar words in this book, especially in the flavor matrix infographics. This is because in certain cases it was advantageous to view a group of ingredients by their higher classifications. Many ingredients have nearly identical flavor compositions, a quirk that allows us to categorize them as a group rather than as individual ingredients. For example, if you can see past the bitterness of one and the sweetness of the other, you will find that cranberries and blueberries have essentially the same flavor. Therefore, throughout this book I'll refer to cranberries and blueberries by their genus, *Vaccinium*. I've done this sort of condensation with the following classifications.

**ALLIUM:** A GROUP OF AROMATIC PLANTS THAT DERIVE THEIR FLAVOR FROM SULFUR COMPOUNDS. THE ALLIUM GROUP INCLUDES: GARLIC, ONIONS, SHALLOTS, LEEKS, SCALLIONS, CHIVES, AND RAMPS. FOR MORE INFORMATION ON ALLIUM, SEE PAGE 18.

**CAPSICUM:** THE ENTIRE CATEGORY OF PEPPERS, FROM MILD BELL PEPPERS TO SPICY CHILES. FOR MORE INFORMATION ON CAPSICUM, SEE PAGE 58.

**SOUTHERN PEA:** ALSO KNOWN AS FIELD PEAS, THIS GROUP INCLUDES BLACK-EYED PEAS, LIMA BEANS, CROWDER PEAS, ZIPPER PEAS, AND MORE. FOR MORE INFORMATION ON SOUTHERN AND OTHER PEAS, SEE PAGE 182.

**VACCINIUM:** THE GENUS OF SHRUBS THAT GROW A VARIETY OF CULINARY BERRIES. BERRIES CLASSIFIED UNDER *VACCINIUM* ARE CRANBERRIES, BLUEBERRIES, BILBERRIES (WHORTLEBERRIES), LINGONBERRIES (COWBERRIES), AND HUCKLEBERRIES. FOR MORE INFORMATION ON VACCINIUM, SEE "BERRY" ON PAGE 42.

Flavor pairing theory can be used to discover wild new combinations like white chocolate and caviar, or to help explain old favorites like pizza. Tomatoes, mozzarella, Parmesan, and baked wheat all share over 100 different aromatic compounds; among them, the pronounced floral aroma of 4-methylpentanoic acid is one of the most significant. All of these shared compounds are the flavor equivalent of an orchestra playing together in perfect time. Each compound has a unique aroma that it contributes to the harmonious flavor of pizza, in much the same way that a cello and flute make different sounds but both contribute to the full, rich sound of a Beethoven symphony.

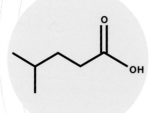

**4-METHYLPENTANOIC ACID COMPOUND**

According to flavor pairing theory, finding the percentages of compounds shared between ingredients would allow me to judge how well certain ingredients pair together relative to others—but generating this percentage was only the next-to-final step in my research process. The last one involved organizing this information in a way that would be quickly and easily understandable—that you could use, at a glance, in your own kitchen.

I have consolidated all of my findings about flavor pairings into what I call a flavor matrix, making one for every ingredient or group of ingredients that I studied. The flavor matrix is an enhanced pie chart that looks a bit like an artist's color wheel. Instead of showing how primary colors can be blended together, however, flavor matrixes show how aromas combine to give each ingredient its unique flavor. Each flavor matrix is essentially the aromatic fingerprint of the ingredient it depicts; it reveals the true aromatic "identity" of that ingredient, and no two ingredients—or matrixes—are exactly the same. Crucially, it also reveals which other ingredients pair best with the central featured ingredient, based on its aromatic identity.

# How to Use This Book

In writing this book, I have been fortunate to have an excellent collaborator: my wife and creative partner, Brooke Parkhurst. Our goal in this book is twofold. We want to teach you about the science of flavor, as we hope we've started to do already. We also want you to experience this revolution for yourself, to master it and put it to use in your own kitchen. In our home, Brooke and I have put flavor pairing theory into practice with the help of hard data about aromatic compounds in food. It has revolutionized the way we cook.

In the pages ahead, we will present you with a series of tools to help you determine which ingredients pair best with one another, based on the science and theory of flavor pairing. Every one of the fifty-eight sections will focus on a different ingredient or group of ingredients, covering roughly 150 common ingredients in all.

The centerpiece of each section is its flavor matrix, an infographic containing all of the data I was able to assemble about a given ingredient's aromatic compounds and those of its compatible ingredients. At the center of each matrix is the featured ingredient. Surrounding that ingredient are its primary aromas and their subcategories, each of which is represented by a "slice" of the inner part of the pie chart. These aromas and their subcategories are explained beginning on page 260.

Around the perimeter of the flavor matrix, extending beyond the inner ring of primary and secondary aromas, additional slices of the pie chart represent all of the ingredients in each aroma group that pair well with the central ingredient. The length of each of these slices indicates the percentage of compounds that the complementary ingredient shares with the featured ingredient on the matrix.

For example, on the olive matrix, you'll see that basil, lovage, and parsley extend farther from the center of the matrix than other herbs in the Vegetal/Herbaceous aroma group, indicating that these ingredients are the best choice to pair with the featured

one, olives. Peaches and nectarine are the biggest slices of this particular matrix, extending farther from the center of the matrix than any other section. This is a reflection of the fact that olives share over 70 percent of their aromatic compounds with the fruits. Indeed, of all ingredients, olives have the strongest affinity for their close botanical relative, tree fruits—a characteristic that is visualized on the flavor matrix.

A quick glance at the olive flavor matrix will give you some ideas about specific ingredients to pair with olives. If you want to mix and match, however, it's worth keeping in mind that some flavor groups go better together than others. To help guide your choices about combining multiple ingredients that all complement the featured ingredient, but which may or may not complement each other, I've created a flavor pairing guide (page 14). This tool will help you create more complex combinations of ingredients by matching three or more ingredients together.

The true value of the flavor matrixes is not simply to show which ingredients pair well. Flavor matrixes can also teach you how to *think* about flavors in food. They will help you to understand ingredients and flavors that complement one another, allowing you to pair foods and create dishes in new and innovative ways. For example, the fact that olives are great with citrus like lemons, orange, and grapefruit also tells us that olives pair well with the *flavor of citrus*—meaning that coriander seeds, passion fruit, Sichuan peppercorns, or lemongrass could be incorporated into dishes with olives if you're all out of lemons—or even if you aren't.

Strong pairings in a flavor matrix, like citrus and olives, signal more than the mere fact that two ingredients will taste good when combined in a dish. It indicates there may be a greater connection between the ingredients, one that is not always apparent. It shows that the ingredients have something in common. Maybe they are native to the same area, or have a botanical relationship or a similar flavor profile.

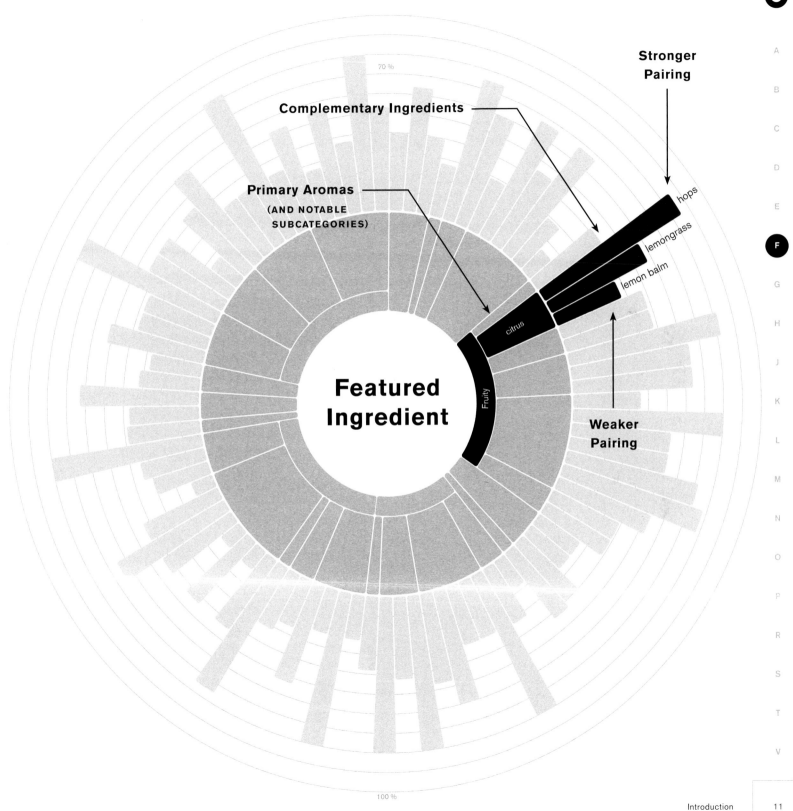

Stronger
Pairing

Complementary Ingredients

Primary Aromas
(AND NOTABLE
SUBCATEGORIES)

hops

lemongrass

lemon balm

citrus

Featured
Ingredient

Fruity

Weaker
Pairing

70 %

100 %

These shared characteristics help to explain flavor compatibility, but they can also help you be more flexible and creative in your use of ingredients. To take just one example, 80 percent of the compounds found in lemongrass are also found in ginger, suggesting they'll taste great when combined in a dish. It also indicates that they have very similar flavor profiles and thus can be easily interchanged in recipes.

Look at these matrixes with a critical eye, in short, and you will begin to understand the way that flavors and tastes are constructed. In short order, this understanding will make you a more creative cook.

Along with the flavor matrix in each section, I list some quick facts about the featured ingredient, including some tips on how to best use it. After that you'll find a Best Pairings list of ingredients that pair especially well with the featured ingredient, as well as a short list of "Surprise Pairings"—combinations that I have rarely come across in my career, even though the ingredients show a strong affinity for one another. You'll also find a list of ingredients that can be substituted for the main ingredient.

Each of the sections in this book ends with an original recipe that we have created using the information in each flavor matrix. For example, in the olive section, you'll find a recipe for using this classic hors d'oeuvre ingredient to make a unique finale for any meal: a dessert of lemon curd with olive oil and crunchy olives. Sweetened with honey and dehydrated in the oven for an extra intense flavor and a delightfully crispy texture, these olives will blow your mind. But even if you only use the recipes for inspiration about how you can incorporate the data in each flavor matrix into your own cooking, our creations will have done their job.

At the back of the book you'll find resources to help satisfy any scientific curiosity you may be feeling after spending time immersed in these flavor matrixes: a listing of the primary aroma categories and subcategories, and suggestions for further reading, as well as the basic tastes and the most significant volatile compounds for each ingredient group. You will also find in-depth pairing charts that focus on some of the more surprising combinations. It notes the specific aromatic compounds that explain these matches, identifies the aromas associated with each compound, and highlights additional ingredients that also contain those compounds. On occasion, we have chosen to use this space to explore some of the most classic pairings, in an effort to understand them better.

To understand how these flavor-pairing charts work, consider the following example: The compounds beta-caryophyllene (aroma of fried, spice, wood) and myrcene (aroma of balsamic, flower, fruit, herb, must) are naturally found in both olives and star anise, helping to explain the strong pairing between these two ingredients. If you look at the olive pairing chart at the back of the book (page 286), you will see that these compounds can also be found in cinnamon, cilantro, fennel, ginger, and turmeric, among other ingredients. So if the olive and star anise pairing intrigues you, you should think about incorporating some of those ingredients into your dish. Of course, you could find these basic pairing suggestions in the flavor matrixes themselves—but by highlighting the molecular commonalities that explain certain pairings, we hope to add another layer of discovery to your use of this book.

Our hope is that once you've gotten comfortable with these flavor matrixes and the science that undergirds them, you'll be able to use this book at a glance to find new, striking ways to use common ingredients in your own cooking. And who knows—when you discover that shared aromatic compounds make (for example) blueberries and horseradish an unexpectedly delicious pairing, and that these same molecules connect horseradish to pork, a roast pork sandwich with blueberry-horseradish jam may become your new favorite meal. See how many different dishes you can create with knowledge like this, and most importantly, enjoy the ride—wherever it takes you.

# How to Pair Tastes

Any seasoned chef will tell you that the most delicious and dynamic dishes result from a combination of tastes. Controlling tastes in food is simple: If a pot of chili tastes flat, add a pinch of salt to draw out the flavors and a dash of hot sauce to kick up the intensity. If the sauce for sautéed salmon is too buttery, a squeeze of lemon juice will cut through the fat.

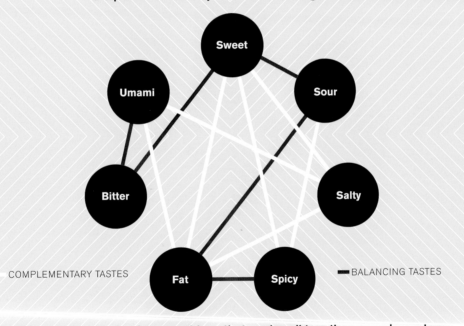

COMPLEMENTARY TASTES          ■ BALANCING TASTES

There are two kinds of taste pairings that work well together: complementary tastes and balancing tastes. Complementary tastes are the kind that accentuate one another; salty and sweet are the best example. Balancing tastes oppose one another to help create harmony; for example, fat mellows out spiciness, and sourness plays down sweetness. The accompanying diagram maps out all of the different complementary and balancing tastes.

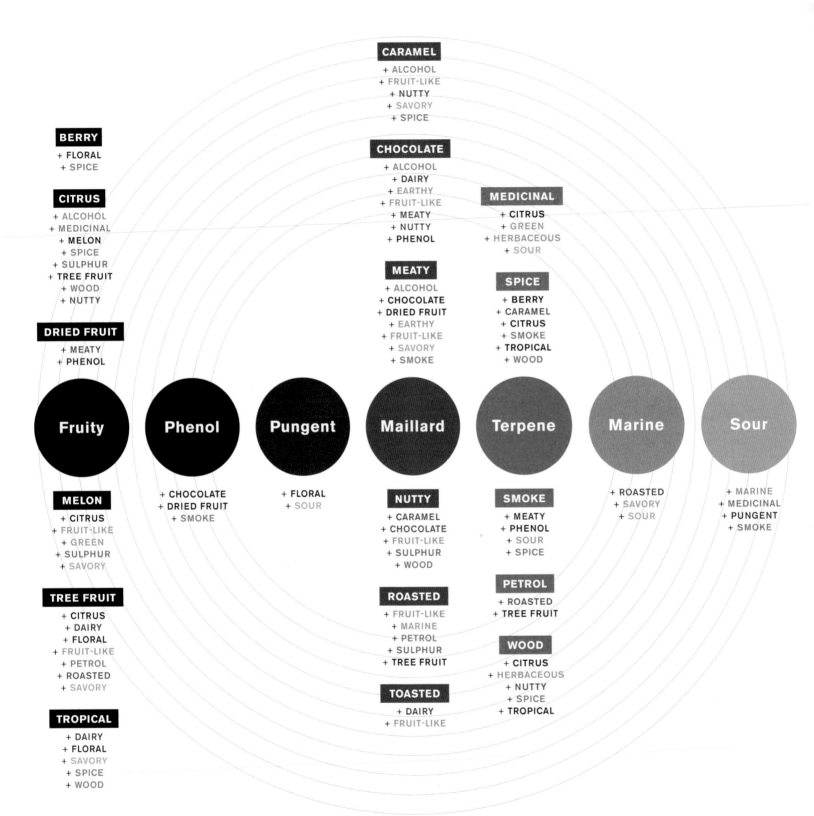

**CARAMEL**
+ ALCOHOL
+ FRUIT-LIKE
+ NUTTY
+ SAVORY
+ SPICE

**CHOCOLATE**
+ ALCOHOL
+ DAIRY
+ EARTHY
+ FRUIT-LIKE
+ MEATY
+ NUTTY
+ PHENOL

**MEDICINAL**
+ CITRUS
+ GREEN
+ HERBACEOUS
+ SOUR

**BERRY**
+ FLORAL
+ SPICE

**CITRUS**
+ ALCOHOL
+ MEDICINAL
+ MELON
+ SPICE
+ SULPHUR
+ TREE FRUIT
+ WOOD
+ NUTTY

**MEATY**
+ ALCOHOL
+ CHOCOLATE
+ DRIED FRUIT
+ EARTHY
+ FRUIT-LIKE
+ SAVORY
+ SMOKE

**SPICE**
+ BERRY
+ CARAMEL
+ CITRUS
+ SMOKE
+ TROPICAL
+ WOOD

**DRIED FRUIT**
+ MEATY
+ PHENOL

**Fruity**  **Phenol**  **Pungent**  **Maillard**  **Terpene**  **Marine**  **Sour**

**MELON**
+ CITRUS
+ FRUIT-LIKE
+ GREEN
+ SULPHUR
+ SAVORY

**TREE FRUIT**
+ CITRUS
+ DAIRY
+ FLORAL
+ FRUIT-LIKE
+ PETROL
+ ROASTED
+ SAVORY

**TROPICAL**
+ DAIRY
+ FLORAL
+ SAVORY
+ SPICE
+ WOOD

+ CHOCOLATE
+ DRIED FRUIT
+ SMOKE

+ FLORAL
+ SOUR

**NUTTY**
+ CARAMEL
+ CHOCOLATE
+ FRUIT-LIKE
+ SULPHUR
+ WOOD

**ROASTED**
+ FRUIT-LIKE
+ MARINE
+ PETROL
+ SULPHUR
+ TREE FRUIT

**TOASTED**
+ DAIRY
+ FRUIT-LIKE

**SMOKE**
+ MEATY
+ PHENOL
+ SOUR
+ SPICE

**PETROL**
+ ROASTED
+ TREE FRUIT

**WOOD**
+ CITRUS
+ HERBACEOUS
+ NUTTY
+ SPICE
+ TROPICAL

+ ROASTED
+ SAVORY
+ SOUR

+ MARINE
+ MEDICINAL
+ PUNGENT
+ SMOKE

# How to Pair Flavors

Whereas flavor matrixes show you flavors in specific ingredients and highlight other ingredients they pair with, this flavor pairing guide will help you to figure out which flavors pair well together. Just as the taste pairing guide on page 13 indicates that sweet and salty tastes make a great match, this guide shows that, for example, citrus and sulfur are complementary flavors.

Here's an example of how you could put the flavor pairing guide to use: In the olive section on page 178, you might look at the flavor matrix and notice that olives and peaches or nectarines are an especially strong pairing.

If you want to build a dish using more than these two ingredients, you'll want to find additional ingredients that pair best with these two. The flavor pairing guide tells us that tree fruit flavors (the flavor group where peaches and nectarines are located) pair nicely with floral flavors. If you have another look at the olive flavor matrix, you'll see some ingredients in the floral section that you could add to your olive–peach/nectarine dish. The results could be any of the following combinations: Olive, Peach, Sichuan Peppercorns, Lavender, and Corn; Olive, Nectarine, Fennel Seeds, Orange, Blossom, and Yogurt; or Olive, Peach, Juniper, Camomile, and Lamb.

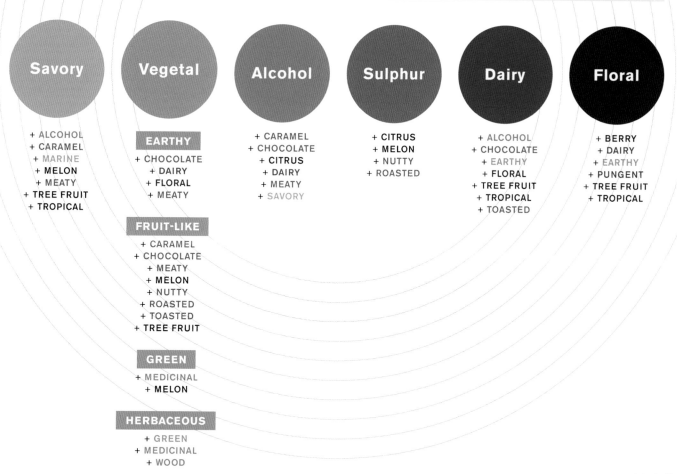

**Savory**
+ ALCOHOL
+ CARAMEL
+ MARINE
+ MELON
+ MEATY
+ TREE FRUIT
+ TROPICAL

**Vegetal**

**EARTHY**
+ CHOCOLATE
+ DAIRY
+ FLORAL
+ MEATY

**FRUIT-LIKE**
+ CARAMEL
+ CHOCOLATE
+ MEATY
+ MELON
+ NUTTY
+ ROASTED
+ TOASTED
+ TREE FRUIT

**GREEN**
+ MEDICINAL
+ MELON

**HERBACEOUS**
+ GREEN
+ MEDICINAL
+ WOOD

**Alcohol**
+ CARAMEL
+ CHOCOLATE
+ CITRUS
+ DAIRY
+ MEATY
+ SAVORY

**Sulphur**
+ CITRUS
+ MELON
+ NUTTY
+ ROASTED

**Dairy**
+ ALCOHOL
+ CHOCOLATE
+ EARTHY
+ FLORAL
+ TREE FRUIT
+ TROPICAL
+ TOASTED

**Floral**
+ BERRY
+ DAIRY
+ EARTHY
+ PUNGENT
+ TREE FRUIT
+ TROPICAL

# INGREDIENTS

**Main Subtypes:**

Garlic, onion, scallion, shallot, leek, ramp, chive

**Best Pairings:**

Brandy, vinegar, wine, citrus, toasted nuts,
cheese

**Surprise Pairings:**

Apple, cocoa, honey

**Substitutes:**

All varieties of allium may be substituted for one
another.

A genus of flowering plants, allium contains many culinary staples. Mainly distinguished by their concentration of sulfur compounds and corresponding pungency, these plants can be ranked accordingly, and as follows: Garlic tends to be most sulfurous and pungent; its bulbs are typically aged for two to three weeks before use, although spring or young garlic (as well as garlic scapes, the green tops of the garlic plant) are milder and can be used fresh. Next comes onion, a bulb whose common forms include yellow onion, stronger-tasting red onion, and milder white onion. Sweet onions—perhaps the mildest of all—are named after the places they are grown (such as Vidalia, Georgia, and Walla Walla, Washington). A separate variety of onion, shallot is a bulb that tends to have a more delicate flavor. Leek is a stalky, broad-leafed cultivar with the mildest flavor of any allium except chives. Ramp, or wild leek, appears only for a short period in the spring in North America. Its flavor is often described as a cross between garlic and scallion. Scallion, or spring onion, is a regular onion variety that is harvested before a bulb forms. Chives have tender green stalks and purple blossoms with a delicate flavor, and which are best used uncooked. With long exposure to heat, all alliums lose their strong flavor and take on a more subtle, roasted, sweet taste.

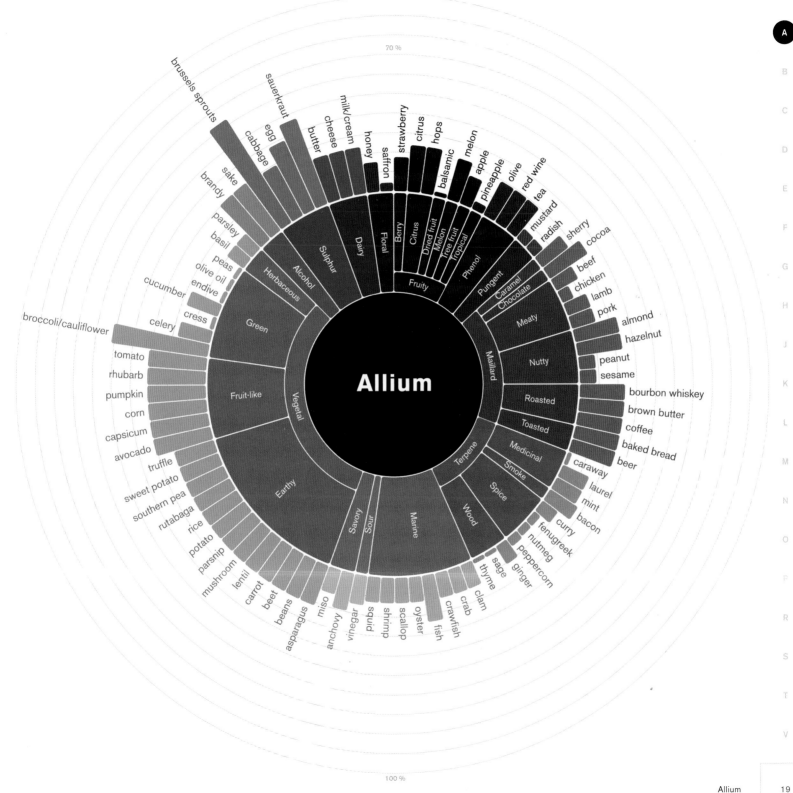

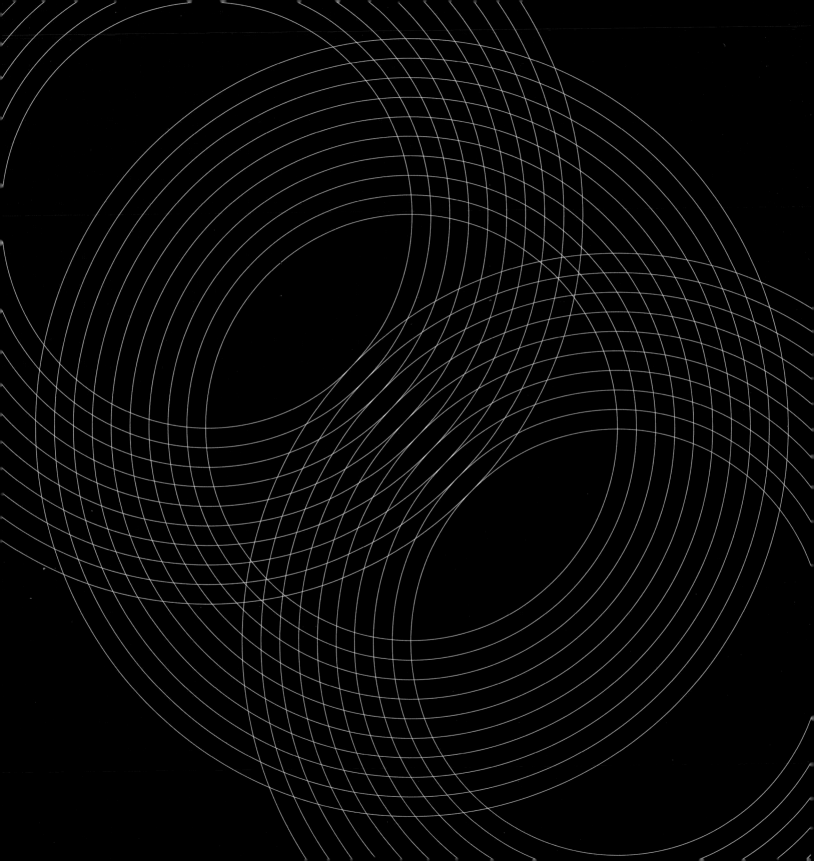

# Garlic Honey

Brooke and I discovered this at the stand of a fantastic honey producer at New York's Union Square farmers' market. It sounded so odd, we just had to grab a sample. When we tasted it, we were blown away. Garlic and honey is an odd combination that has been receiving more attention lately. The pairing makes sense from a chemical perspective: The strong sulfur aromas of garlic are also prominent in honey. In this recipe, those pungent aromas get punched up with dried chile and mustard powder. Drizzle over cheese and bread, or mix into marinades for vegetables or meat.

½ cup raw honey
1 tablespoon minced garlic
½ teaspoon red pepper flakes
½ teaspoon dry mustard powder

Place the honey in a small jar, tighten the lid, and place the jar in a bowl of hot tap water for 10 minutes. (This will melt the honey slightly and make mixing easier.) Stir in the garlic and pepper flakes. Replace the lid. Store at room temperature for 3 to 5 days before using.

**MAKES ABOUT ½ CUP**

The flavor of cooked artichokes is dominated by earthy aromas, although sulfurous, herbal, and citrus aromas contribute to this plant's flavor profile as well. Artichokes have an interesting ability to make other foods taste sweeter, because cynarin—a compound naturally present in the plant—inhibits our taste buds' ability to perceive sweetness; when a bite of a different ingredient washes away the artichoke's cynarin, your brain interprets the taste buds' reactivation as a mild increase of sweetness.

**Best Pairings:**

Citrus, mint, basil, mushroom, sesame, wine, cheese, cocoa

**Surprise Pairings:**

Plum, yogurt, sesame, cilantro

**Substitutes:**

Jerusalem artichoke, hearts of palm, chayote, asparagus

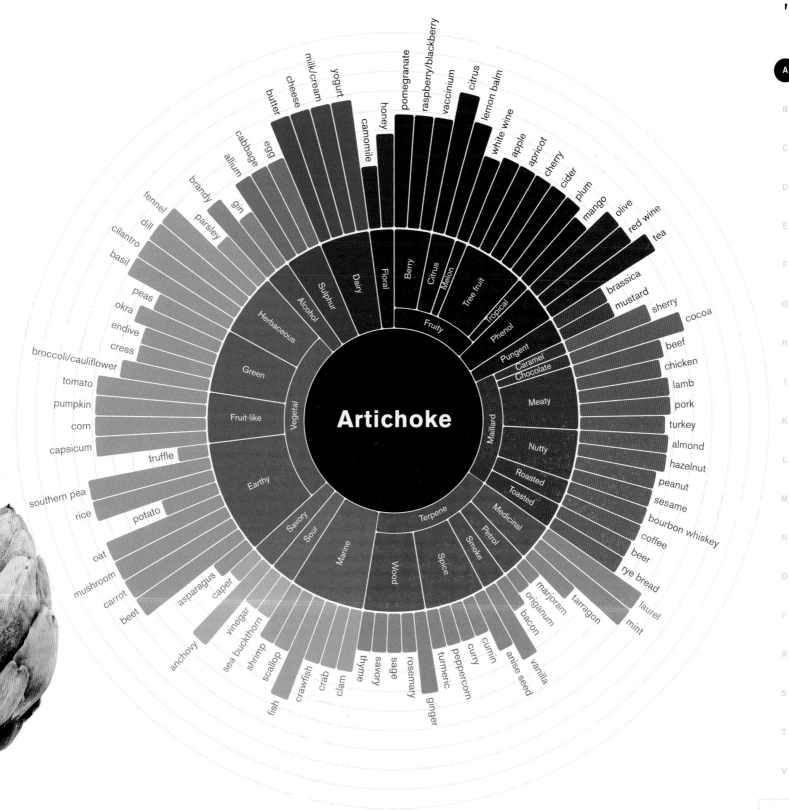

Artichoke

# Tahini Vinaigrette

Inspired by the artichoke's unique chemical makeup, this dressing is a balance of toasted, savory, and tart flavors. It's the perfect complement to artichokes, but makes an equally delicious sauce for asparagus and seafood, especially grilled fish, shrimp, and clams.

1 lemon
½ teaspoon fennel seeds
1 teaspoon coriander seeds
½ teaspoon black peppercorns
½ cup extra-virgin olive oil
2 sprigs fresh thyme
1 clove garlic, smashed
2 tablespoons tahini (sesame seed paste)
1 teaspoon minced jalapeño chile or ½ teaspoon chili paste
Kosher salt

Remove the zest from the lemon in small strips with a vegetable peeler. Juice the lemon. Set the juice and zest aside separately.

Combine the fennel seeds, coriander seeds, and black pepper in a mortar and crush lightly with the pestle. Or place the spices between two sheets of parchment paper and crush them with the bottom of a small pot. Transfer the spices to a small dry sauté pan and toast them over medium-low heat until very aromatic. Do not let them begin to smoke. Remove the pan from the heat and add the olive oil, thyme, garlic, and lemon zest. Return the pan to medium-low heat and cook just until the garlic begins to sizzle. Remove from the heat, cover, and set aside 15 minutes for the flavors to infuse the oil.

In a small bowl, stir together the lemon juice, tahini, and jalapeño. Strain in the cooled oil and whisk to combine. Season to taste with salt. Serve immediately, or transfer to a glass jar or plastic container, seal tightly, and store in the refrigerator for up to 14 days.

**MAKES ABOUT 1 CUP**

Asparagus, commonly thought of as a green vegetable, is actually related to alliums. Accordingly, asparagus's flavor comes mostly from earthy and sulfur aromas. The sulfur aromas in asparagus become stronger as it is cooked, which is why it is very important to avoid overcooking. You can find asparagus in green, purple, and white varieties. Green and white are actually the same plant; the white kind is simply grown in the absence of light, to prevent the plant from developing chlorophyll. Growers accomplish this by mounding hay or dirt over the rows of asparagus to block sunlight. These tougher growing conditions also create the thicker stalks that are typical of white asparagus. Purple asparagus is a separate cultivar that tends to be sweeter and less fibrous. Be aware, however, that the unique color of purple asparagus turns green when cooked.

**Best Pairings:**

Beer, bread, seafood, butter, cheese, egg, ham, vinegar

**Surprise Pairings:**

Beer, coconut, potato chips, fish sauce

**Substitutes:**

Artichoke, broccoli, cauliflower (especially thinly sliced stems)

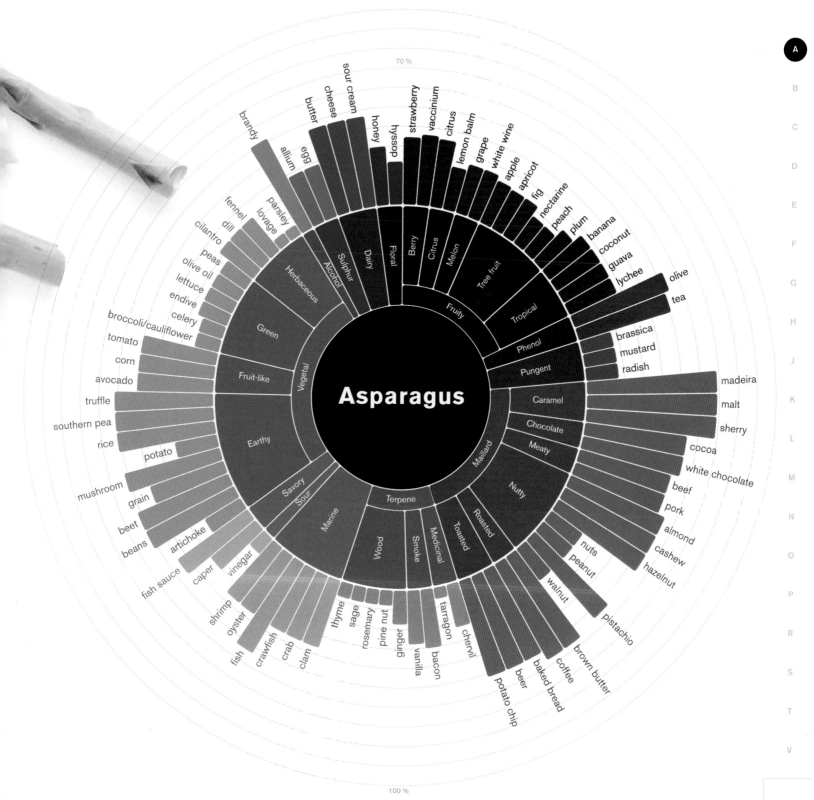

# Beer-Pickled Mustard Seeds

Asparagus is notoriously difficult to pair with wine, but its chemical affinity for beer is off the charts. Here, we combine two of asparagus's favorite things—mustard and beer—into a condiment that pairs perfectly with any asparagus dish. Serve on its own with steamed asparagus as an hors d'oeuvre or as a topping for grilled or roasted asparagus. Try mixing it into sour cream or crème fraîche to dollop on cooked asparagus, or use to finish sauces paired with asparagus.

½ cup yellow mustard seeds
¼ cup brown mustard seeds
1 cup apple cider vinegar
2 tablespoons light brown sugar
1 teaspoon kosher salt
½ cup pale ale
Cayenne pepper, piment d'espelette, or crushed
    peperoncino (optional)

Combine the mustard seeds in a small saucepot and cover completely with cold water. Bring to a boil. Remove from the heat immediately and drain in a fine-mesh sieve. Rinse the mustard seeds thoroughly with cold water. Return to the pot and repeat the process twice, boiling and rinsing a total of three times.

Return the mustard seeds to the pot and add the vinegar, sugar, and salt. Bring to a simmer, then remove from the heat and set aside to cool. Stir occasionally to prevent the seeds from sticking to the bottom of the pot. When cooled slightly, stir in the ale. For a spicier mustard, stir in cayenne to taste.

Transfer the mustard to a glass jar or plastic container, seal tightly, and refrigerate for 24 hours before using.

**MAKES APPROXIMATELY 2 CUPS**

Avocado trees are botanically related to laurel (bay) and cinnamon trees, which helps to explain why these ingredients show such affinities for one another. But avocadoes have some unique properties as well. For example, their particular flavor is created by a high concentration of monounsaturated fats, lipids that give the fruit (yes, the avocado is a fruit!) its characteristic aromas of green, dairy, citrus, and floral. As the fats oxidize through cooking or ripening, they create Maillard aromas. Native to Mexico, avocadoes are grown in tropical and Mediterranean climates throughout the world. They may be harvested year-round, thanks in part to the fact that they ripen after being picked.

**Best Pairings:**

Cocoa, chiles, fruit (especially citrus), butter, cream, roasted meat, seafood

**Surprise Pairings:**

Tequila, cocoa, apple, eggplant

**Substitutes:**

Chayote, artichoke, mashed peas or beans

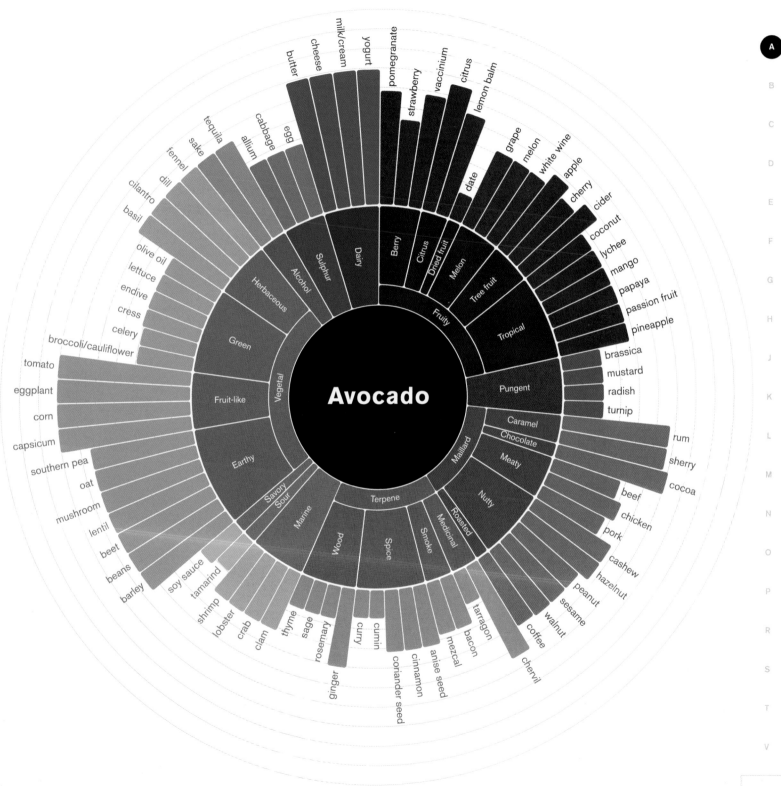

Avocado

# Sesame Seed and Avocado Salad with Fig Vinegar

Classic pairings for avocado—chile and lime—meet some unexpected new friends in this dish. Fig plays up the avocado's natural fruitiness, so don't overlook the fig vinegar; it's a hidden gem in this recipe and a flavorful secret weapon to keep in your pantry. Serve this salad on its own, with the Cocoa and Chile–Rubbed Beef on page 37, or with grilled shrimp.

2 avocadoes, cut in half, pitted, and peeled
1 lime
Cocoa and Chile Rub (page 37)
¼ cup white sesame seeds, toasted
4 radishes, trimmed and thinly sliced
2 cups baby arugula
2 tablespoons extra-virgin olive oil
Kosher salt
Fig Vinegar (recipe follows)

Cut the avocado into wedges, 4 pieces per half. Squeeze a bit of lime juice over the avocado wedges, then season them with the cocoa-chile mixture. Let the wedges stand for 1 minute.

Spread the sesame seeds on a shallow plate. Dip one cut side of each wedge into the sesame seeds, pressing to make sure they stick. Set the wedges aside on another plate.

Toss the radishes and arugula with the olive oil and salt to taste. Divide among four plates and top with the avocado wedges. Drizzle each plate with fig vinegar.

## FIG VINEGAR

¼ cup chopped dried figs
1 cup water
2 cups white wine vinegar

Combine the figs and water in a small saucepot. Bring to a simmer and cook until the pot looks nearly dry. Add the vinegar and return to a simmer. Remove the pot from the heat and cover tightly. Set aside for 30 minutes to infuse.

Mash the figs with a fork, then press the mixture through fine-mesh sieve. (Discard the seeds and any remaining pulp.) Transfer the vinegar to a glass jar and seal tightly. Store in a cool, dark place for up to 6 months.

**Makes 2 cups**

**MAKES 4 SERVINGS**

A
B
C
D
E
F
G
H
J
K
L
M
N
O
P
R
S
T
V

Beef's flavor is determined in part by the breed of cattle from which it comes, and to an even greater extent by the animal's diet and fatty acids derived from the animal's diet. Specifically, linolenic, oleic, and caproic acids are the main contributors to flavor in raw beef. Linolenic acid gives beef a clean fatty taste; oleic and linolenic acids both provide a fatty, fried flavor; and caproic acid is desirable in beef for the cheesy-fatty flavor it provides. Grass-fed cattle have higher concentrations of linolenic acid in their flesh; grain-fed cattle develop more oleic and linoleic acid; corn-finished cattle, meanwhile, develop less caproic acid. Aging also plays a significant role in beef flavor; beef has very little flavor when freshly slaughtered, so retail cuts of meat typically undergo a minimum of seven days aging—the time it takes for beef's distinctive "meaty" flavor to begin to develop. This process has a chemical effect: After fourteen days, the aromatic compound 1-octen-3-ol, a key part of beef's flavor, can increase in the meat by more than 1,000 percent.

**Best Pairings:**

Nuts, dried fruit, butter, cream, mustard, alliums, cocoa

**Surprise Pairings:**

Cocoa, grapes, dried currant

**Substitutes:**

Venison, lamb, bison, tuna

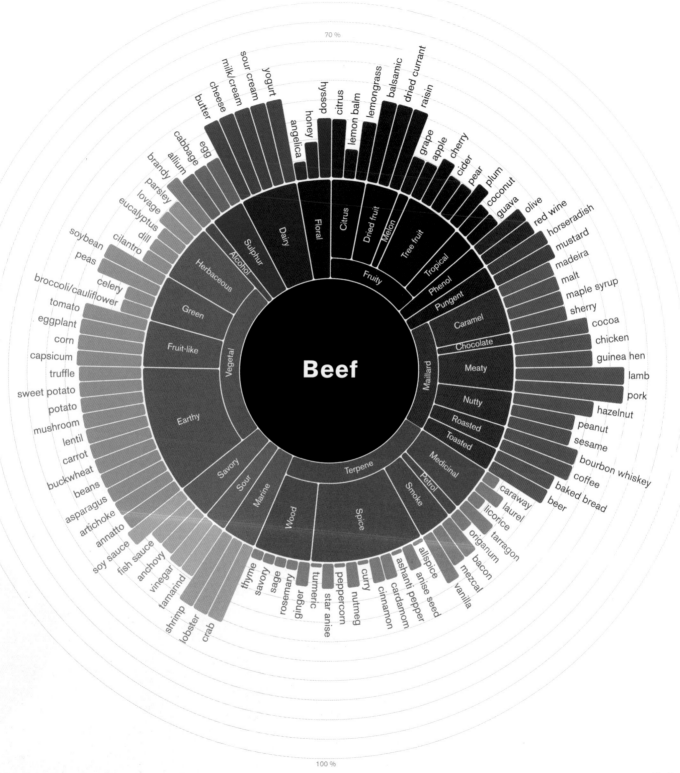

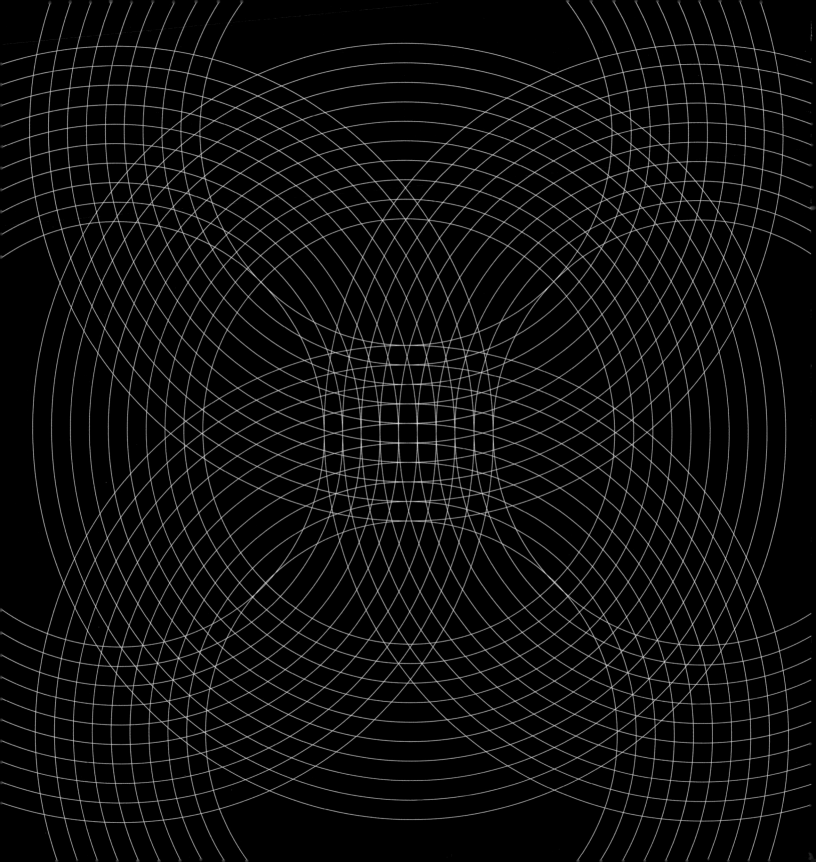

# Cocoa and Chile–Rubbed Beef

Like most meats, beef derives a lot of flavor from the Maillard reactions that take place during cooking. In this recipe, cocoa and dried chiles augment those roasted aromas, packing even more flavor into the beef. Serve with Sesame Seed and Avocado Salad with Fig Vinegar (page 33).

## COCOA AND CHILE RUB

1 tablespoon unsweetened cocoa powder

1 tablespoon kosher salt

2 teaspoons ancho chile powder

1 teaspoon sesame seeds (optional, but excellent)

1 teaspoon freshly ground black pepper

2 pounds flank steak, trimmed of excess fat

1 tablespoon canola oil

*Make the rub:* Combine the cocoa, salt, ancho powder, sesame seeds (if using), and pepper in a small bowl and mix well.

Season the steak on both sides with the rub. You may not use all of it. Store any leftover rub in a tightly sealed container in a cool, dry place.

Position a wire rack over a baking sheet. Heat the canola oil in a large sauté pan over medium-high heat and sear the steak on both sides until golden brown, 4 minutes per side for medium-rare. Or heat a grill for direct grilling until hot and cook for 3 to 4 minutes per side. Cook more or less if you prefer a different doneness. Transfer to the rack to rest for at least 4 minutes. Slice into thin pieces, cutting against the grain.

**SERVES 4**

Table beets—as opposed to sugar beets, which are grown to make sugar—have a variety of culinary uses. Like most root vegetables, they may be eaten either raw or cooked. Roasting brings out the natural sugars in beets, giving them a very pleasant earthy-sweet flavor. (Geosmin, the aroma compound most responsible for beets' earthy flavor, is also responsible for the earthy aroma that fills the air when rain falls on dry soil.) Although beet leaves are eaten less often, they too have excellent flavor. Young leaves may be added to salads or eaten raw; larger leaves may be sautéed like spinach or other greens or chopped and added to soups and stews, just like chard, which is also a member of the beet family.

**Best Pairings:**

Citrus, toasted nuts, cheese, yogurt, seafood, wine, vinegar

**Surprise Pairings:**

Lychee, oyster, lemon balm, tea, apple

**Substitutes:**

Carrot, parsnip, cabbage

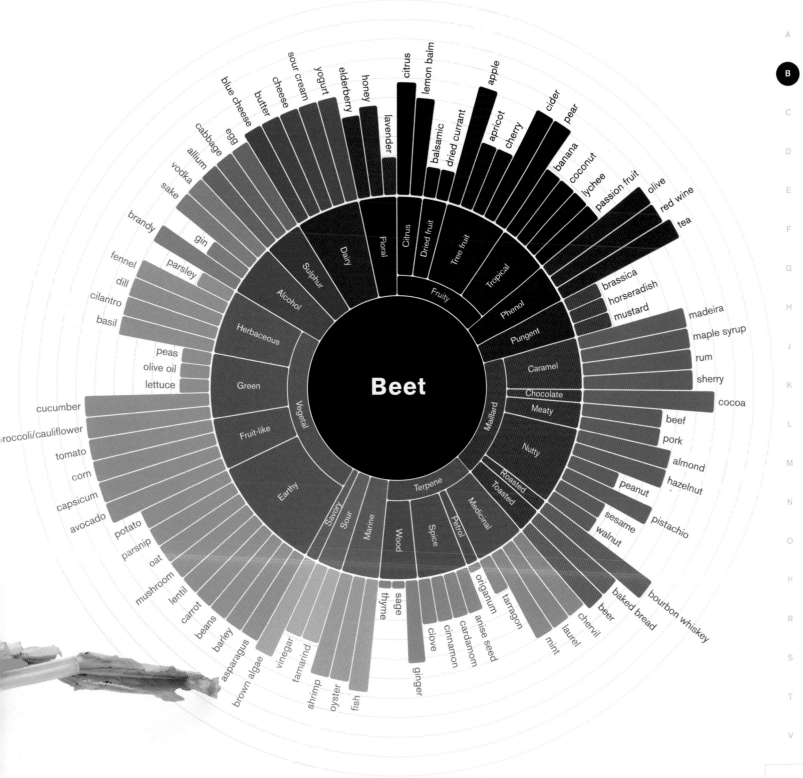

Beet

# Beet and Carrot Salad
# with Orange-Hazelnut Vinaigrette

Beets show a particular affinity for the earthy flavor of fellow root vegetables, like carrots and jicama, as well as ingredients with fruit-like aromas. The unique flavors and textures in this salad are also inspired by one of our favorite dishes on the menu at UXAU Casa Hotel & Spa in Trancoso, Brazil.

2 cups peeled and shredded raw red beets

2 cups shredded carrots

2 cups peeled and shredded jicama or shredded green
  cabbage

4 scallions (green and white parts), thinly sliced

Orange-Hazelnut Vinaigrette (recipe follows)

Kosher salt and freshly ground black pepper

½ cup chopped toasted cashews

1 tablespoon chopped parsley

Combine the beets, carrots, jicama, and scallions in a large bowl. Add orange-hazelnut vinaigrette as needed (about ¼ cup). Toss well and season to taste with salt and pepper. Just before serving, mix in the cashews and parsley.

## ORANGE-HAZELNUT VINAIGRETTE

¼ cup white wine vinegar

1 tablespoon minced shallot

Kosher salt and freshly ground black pepper

2 tablespoons finely chopped hazelnuts

½ cup canola or vegetable oil

1 teaspoon grated orange zest

2 tablespoons fresh orange juice

1 teaspoon Dijon mustard

¼ cup extra-virgin olive oil

Combine the vinegar, shallot, and a pinch each of salt and pepper in a medium bowl. Mix well. Set aside to macerate while preparing the hazelnut oil.

Combine the hazelnuts and canola oil in a small sauté pan and place over medium heat. When the nuts begin to bubble in the oil, remove from the heat and set aside until cooled to room temperature.

Whisk the orange zest, orange juice, and mustard into the vinegar and shallot. Slowly pour the hazelnut and olive oils, whisking vigorously. The mixture should thicken and emulsify. Use immediately. Store any leftover vinaigrette in a tightly sealed glass jar in the refrigerator for up to 3 weeks; shake or whisk to recombine before using.

**Makes about 1 cup**

**SERVES 4**

Berry is a wide category: Botanically speaking, it covers fruits—that is, the ripened ovary of a plant—that enclose seeds. By this definition, bananas, avocadoes, tomatoes, eggplants, watermelons, and pumpkins are all berries, but strawberries, raspberries, blackberries, and such are not; they are aggregate fruits, each made up of many smaller fruits. I will focus here on popularly defined berries like strawberries and blackberries. Grouping these fruits into three main genera helps us to focus on the similarities among all berries and identify the particular characteristics in each genus. Fragaria is a genus of the rose family that produces strawberries, including cultivars and varieties such as Honeoye, Earliglow, Allstar, and fraises des bois. Rubus is a separate genus of the rose family that produces raspberries, blackberries, dewberries, loganberries, and boysenberries. Vaccinium is a genus in the heath family that includes blueberries, cranberries, and lingonberries. (The species in this genus are often referred to in this book by their genus, and sometimes individually.) Note that when it comes to pairing berries with other ingredients, all three groups have similar affinities.

**Main Subtypes:**

Strawberry, blueberry, blackberry, raspberry, cranberry

**Best Pairings:**

Citrus, melon, apricot, peach/nectarine, chocolate, arugula, wine, vinegar, cream, yogurt

**Surprise Pairings:**

Basil, mushroom, cumin, olive

**Substitutes:**

Other berries, currants, grapes, kiwi, pomegranate

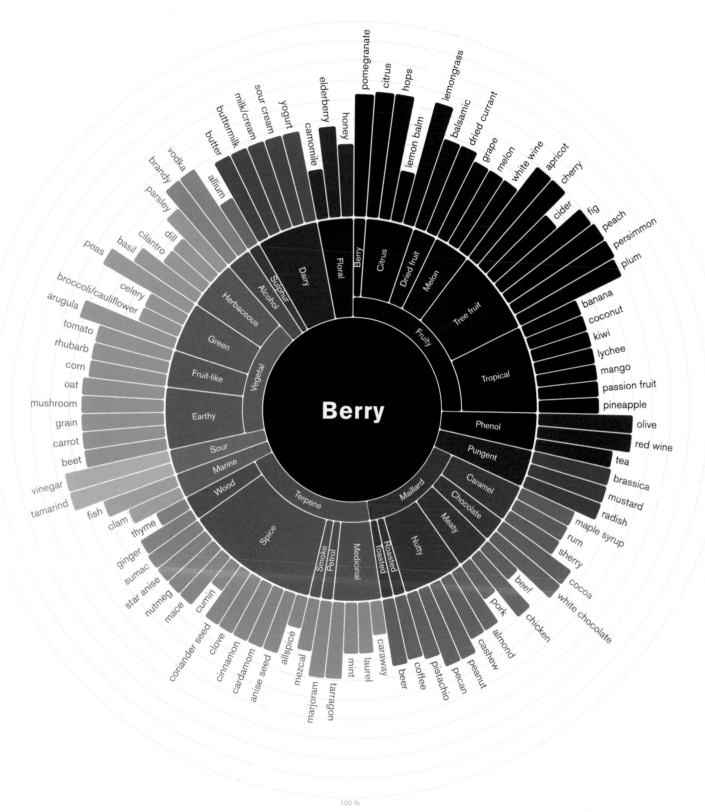

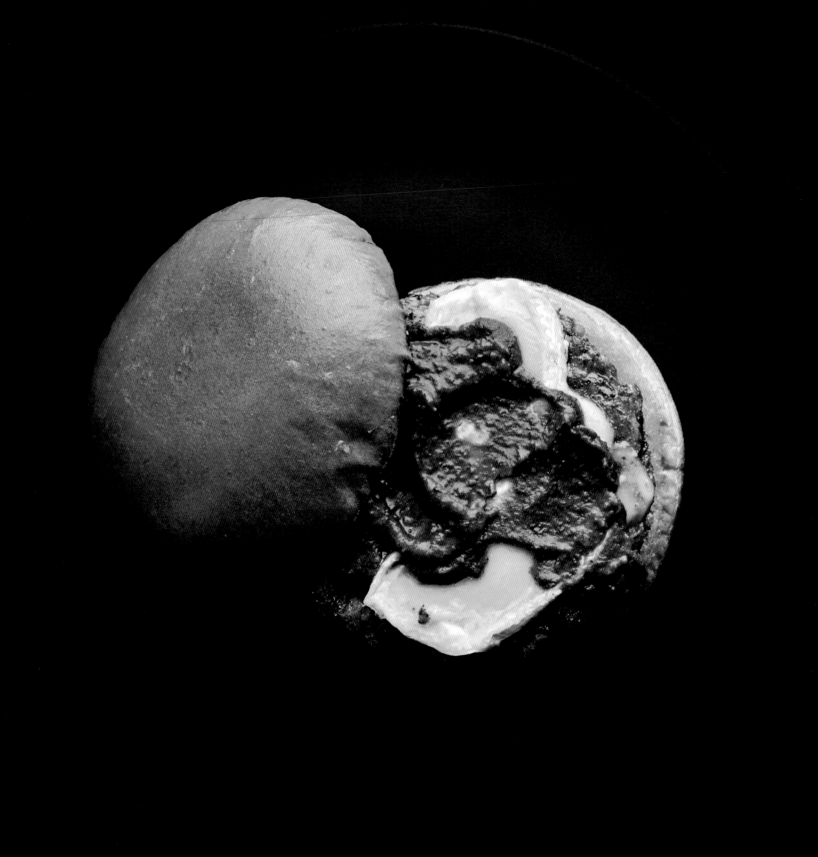

# Chicken and Mushroom Burgers
# with Strawberry "Ketchup"

This pairing has traveled around the world. Combining chicken, mushrooms, and strawberries was first conceived by Michael Laiskonis and Florian Pinel with IBM's Chef Watson in New York City. Michael brought his take on the pairing to SXSW in Austin, Texas. Later, my take (the below recipe) became the subject of a talk I gave at TEDxWarsaw. Now it lands in your kitchen.

2 cloves garlic, chopped
1 shallot, minced
2 tablespoons olive oil, plus more for cooking the burgers
8 ounces fresh mushrooms, coarsely chopped
1 pound ground chicken
6 tablespoons dry breadcrumbs
2 large eggs, beaten
2 tablespoons chopped fresh basil
1 teaspoon kosher salt
½ teaspoon freshly ground black pepper
5 slices mild white cheese (Brie, provolone, or Muenster)
5 burger rolls, split
Softened butter, for the rolls
Strawberry "Ketchup" (recipe follows), for serving

Sauté the garlic and shallots in the olive oil over medium heat until softened. Add the mushrooms and cook until browned. Set aside on a plate to cool.

Combine the cooled mushroom mixture, chicken, breadcrumbs, eggs, basil, salt, and pepper in a bowl. Blend well and form into 5 burgers.

Place a sauté pan with a thin layer of oil over medium-high heat. When hot, pan-fry the burgers for about 6 minutes per side, flipping once, until well browned and cooked through (the internal temperature 165°F). Immediately top each hot burger with a cheese slice. If necessary, transfer to a 425°F oven to finish cooking. Set aside to rest while you prepare the rolls.

Heat the broiler. Spread the cut sides of the rolls with butter and broil until they are toasted. Place the burgers on the bottoms of the rolls, top with strawberry ketchup, and serve.

## STRAWBERRY "KETCHUP"

2 tablespoons minced shallots
2 tablespoons minced peeled fresh ginger
2 cloves garlic, minced
1 tablespoon unsalted butter
2 cups chopped mushrooms (about 5 ounces)
Kosher salt
4 cups chopped fresh strawberries (about 1 pound)
1 teaspoon crushed coriander seeds
Freshly ground black pepper
Sugar (optional)

Sauté the shallots, ginger, and garlic in the butter until tender and fragrant. Add the mushrooms, season lightly with salt, and cook until completely dry, about 6 minutes.

Stir in the strawberries, coriander, and ½ teaspoon pepper. Cook over medium heat, stirring occasionally, until the strawberries become very soft, about 10 minutes. Blend in the pot with an immersion blender, or transfer to a blender and process until smooth, then return to the pot. Simmer until thickened, about 3 minutes more, stirring occasionally to prevent the ketchup from burning on the bottom.

Season to taste with salt, pepper, and a pinch of sugar (if necessary). Transfer to a glass jar or plastic container and let cool to room temperature. Seal tightly and store in the refrigerator for up to 2 weeks.

**Makes 3 cups**

**MAKES 5 BURGERS**

**Main Subtypes:**

Broccoli, cauliflower, romanesco

**Best Pairings:**

Citrus (especially lemon), egg, milk, cream, cheese, cocoa, curry, cilantro, fish sauce

**Surprise Pairings:**

Peanut, fig, cocoa

**Substitutes:**

Broccoli, cauliflower, and romanesco can substitute for each other. Also: asparagus, Brussels sprouts, broccoli rabe, kale

Broccoli, cauliflower, and romanesco are members of the *Brassica oleracea* species. But have their own unique characteristics that distinguish them from their leafier cousins, mainly their colors and the shape of their flowering heads. The different shapes of each of these three varieties affect both their appearance and texture after cooking. The tightly formed heads of cauliflower and romanesco result in a firmer texture when cooked, while the higher water content of broccoli requires care- ful cooking to prevent it from becoming mushy. All three derive their flavor from sulfur compounds, as do other brassicas—although these three have stronger green, vegetal, and citrus aromas.

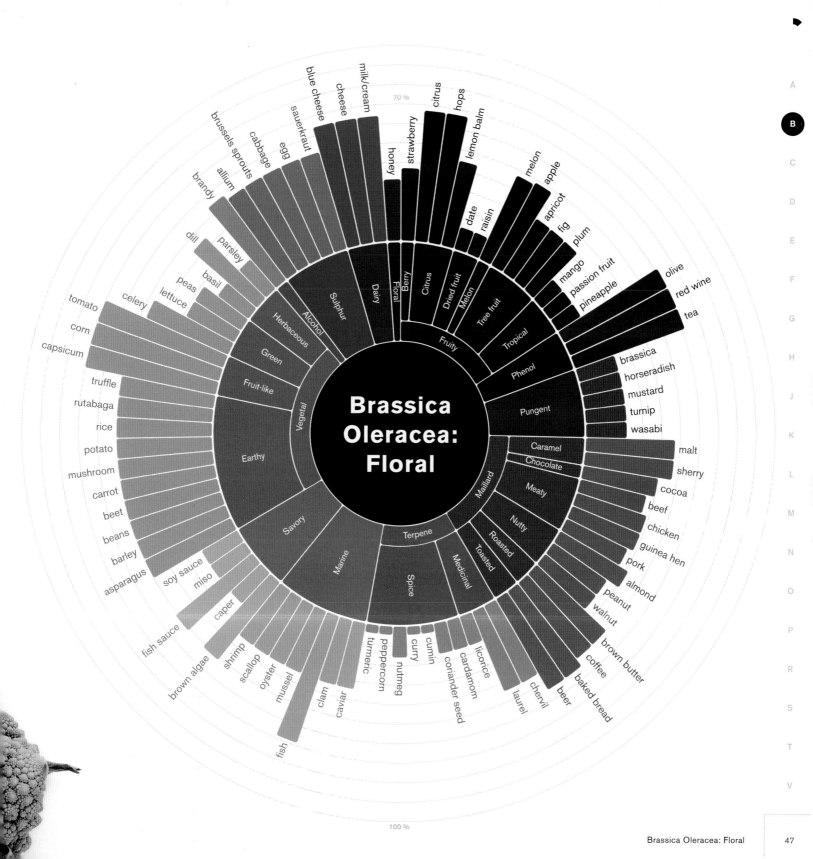

# Brassica Oleracea: Floral

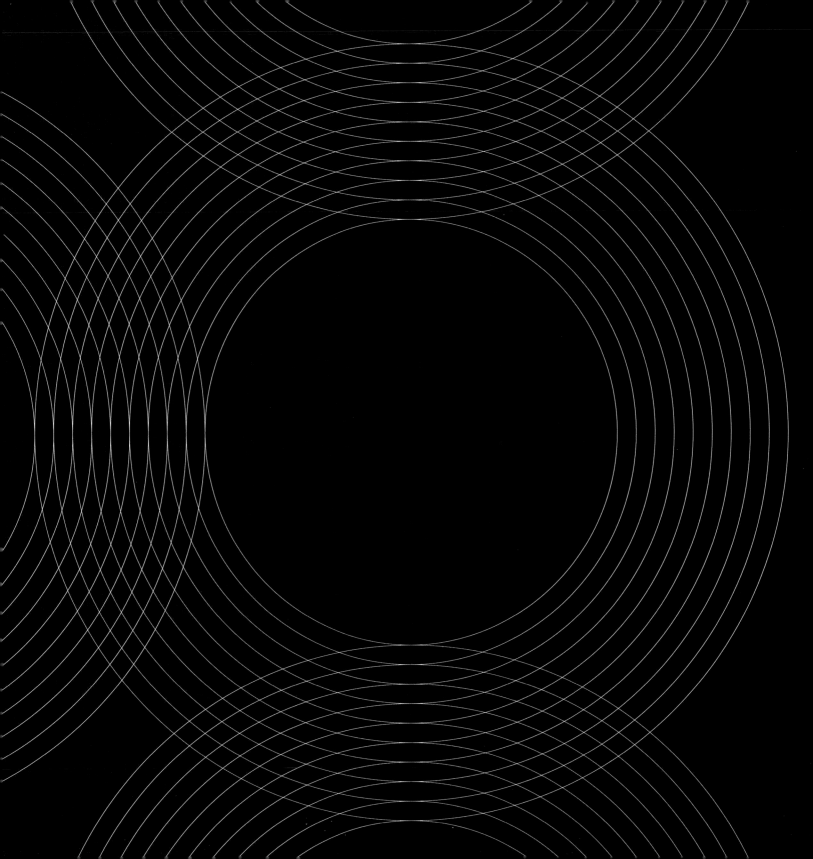

# Broccoli, Peanut, and Pumpkin Stir-Fry

Fans of Thai cooking may already be comfortable with the pairing of broccoli, peanuts, and fish sauce, but pumpkin rarely gets invited to the party. For an even more unique pairing, try serving this dish with grilled clams or oysters.

## VEGETABLES

2 bunches broccoli (about 1½ pounds)
1 small pumpkin or large acorn squash (about 1½ pounds)
4 cloves garlic, peeled and smashed
2 tablespoons fish sauce
3 tablespoons vegetable oil

## SAUCE

1 cup chicken stock
2 tablespoons fish sauce
1 tablespoon lime juice
1 tablespoon honey
1 tablespoon cornstarch

## STIR-FRY

2 tablespoons vegetable oil
3 cloves garlic, sliced
6 scallions, trimmed and sliced into ½-inch pieces
2-inch piece fresh ginger, peeled and thinly sliced
2 whole dried arbol chiles
2 tablespoons cilantro leaves
¼ cup toasted peanuts, chopped

Preheat the oven to 450°F.

*Prepare the vegetables:* Separate the heads and stems of the broccoli. Peel and thinly slice the stems; cut the heads into bite-sized florets. Set all the broccoli pieces aside in a large bowl. Peel the pumpkin and halve it from top to bottom. Scoop out the seeds and fibers. Thinly slice the pumpkin flesh. Add to the broccoli. Add the garlic, fish sauce, and oil and toss well to coat. Spread the vegetables in a single layer on a large baking sheet or in roasting pan. Use two pans if needed; it is important the vegetables not be crowded.

Roast the vegetables for about 12 minutes, until they are lightly browned at the edges and easily pierced through with the tip of a small knife. Set aside to cool.

*Make the sauce:* Combine the sauce ingredients in a bowl and whisk well to combine. Set aside.

*Make the stir-fry:* Heat a large sauté pan over high heat. Add the oil, garlic, scallions, ginger, and chiles and stir constantly to prevent browning. When very aromatic, add the roasted vegetables. Toss well and continue stirring. When the mixture begins to sizzle rapidly, add the sauce. Stir well until the sauce boils and thickens.

Divide among four plates and garnish with cilantro and peanuts.

**SERVES 4**

**Main Subtypes:**

Cabbage, Brussels sprout, kohlrabi, kale, collard greens

**Best Pairings:**

Toasted nuts, cheese, mushroom, grains, anchovies

**Surprise Pairings:**

Coconut, cilantro, eggplant

**Substitutes:**

Any *Brassica oleracea* or *B. rapa* may be substituted for another, including broccoli and cauliflower

Like most brassicas, the leafy members of the *oleracea* species—cabbage, Brussels sprouts, kohlrabi, kale, and collard greens—derive the majority of their flavor from sulfur compounds. The distinctive smell of boiled cabbage or Brussels sprouts is due to the release of sulfur compounds during cooking. Green aromas are also found throughout the species, while the dark green leafy varieties (kale, collard greens) contain bitter phenolic compounds as well.

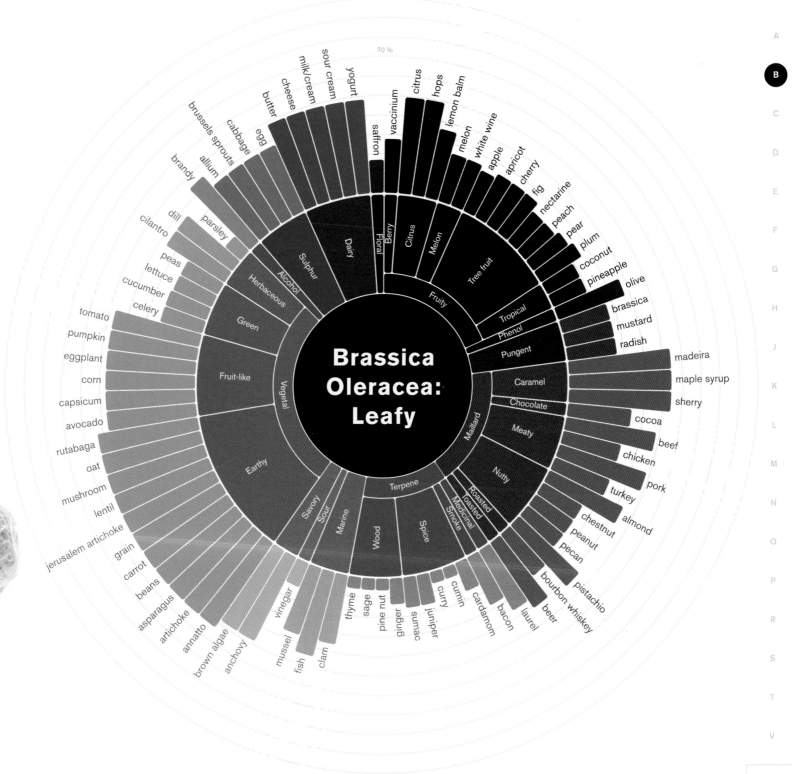

Brassica Oleracea: Leafy

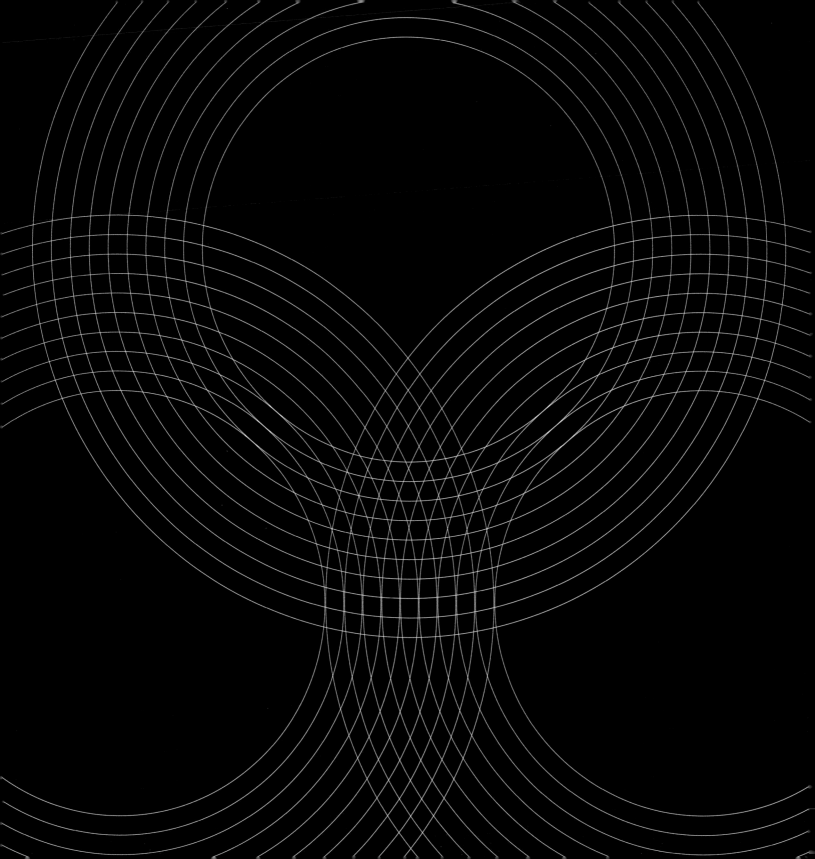

# Umami Vinaigrette

I love using this dressing for salads that include kale, cabbage, shaved Brussels sprouts, or other brassicas. It's also perfect for coating Brussels sprouts, broccoli, or cabbage before roasting.

½ cup red wine vinegar
1 tablespoon chopped shallot
2 cloves garlic, peeled
4 anchovy fillets
1 tablespoon Dijon mustard
½ teaspoon mushroom powder (optional)
¼ cup grated Parmesan cheese
1 cup canola or vegetable oil
½ cup extra-virgin olive oil

Combine the vinegar, shallot, garlic, anchovies, mustard, mushroom powder (if using), and Parmesan in a blender. Blend on low speed until smooth. Slowly add the oils with the machine running. When all the oil has been added, the vinaigrette should be emulsified. Store in a tightly sealed glass jar in the refrigerator for up to 4 weeks. Shake or whisk to recombine before using.

**MAKES ABOUT 2 CUPS**

*Brassica rapa* is a group of plants with a number of subspecies, from turnips to bok choy, that are grown for both culinary uses and oil production. (The seeds of *Brassica rapa* plants are pressed to create canola oil.) When cooked, most *Brassica rapa* have a flavor dominated by strong-smelling sulfur compounds. Among these compounds is allyl isothiocyanate, which is also found in radishes, horseradish, and mustard, and which has a spiciness similar to that of the compound capsaicin, found in hot peppers.

**Main Subtypes:**

Napa cabbage, bok choy, broccoli rabe, turnip, mustard greens

**Best Pairings:**

Berries, butter, cream, cheeses, alliums

**Surprise Pairings:**

Melon, camomile, hazelnut

**Substitutes:**

Spinach, kale, broccoli, cauliflower

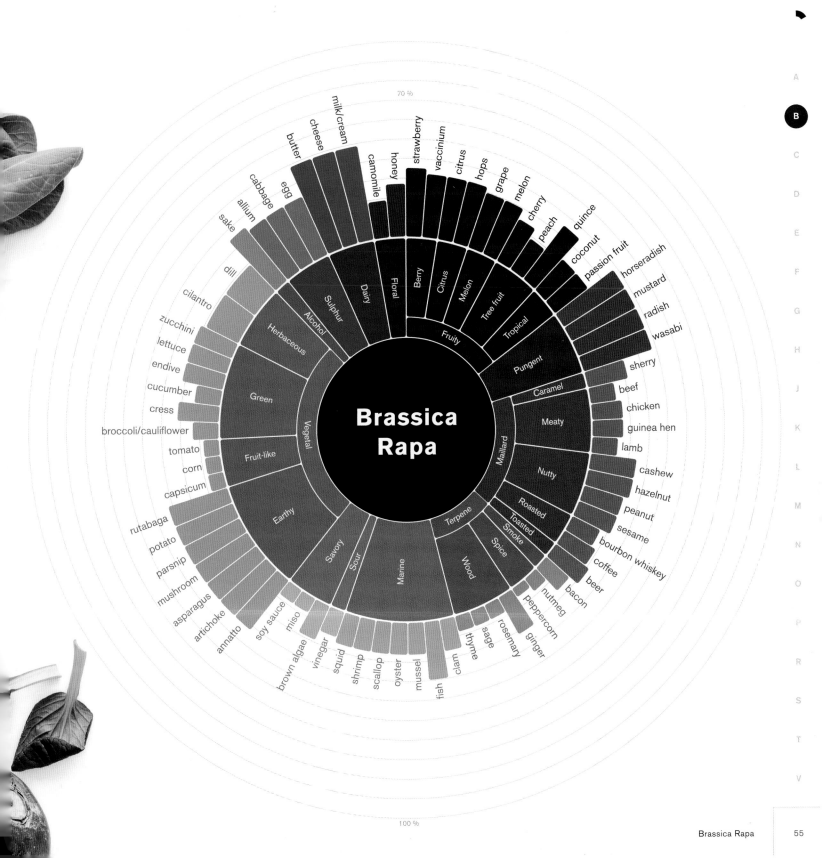

# Brassica Rapa

# Pan Roast of Turnips, Apple, Chicken, and Radishes with Hazelnuts

This is one of those lovely meals that comes together in just about 30 minutes and leaves you with only one pan to clean. As if that weren't enough to convince you, it packs an outrageous combination of flavor and textures, and is just as delicious without the chicken and served as a vegetable side.

8 boneless, skinless chicken thighs
2 teaspoons kosher salt
6 branches fresh thyme, plus 1 teaspoon leaves
1 teaspoon dry mustard
½ teaspoon freshly ground black pepper
2 tablespoon canola or vegetable oil
3 tablespoons unsalted butter
8 baby turnips, trimmed and cut in half
1 Golden Delicious apple, peeled, cored, and cut into
    8 wedges
8 radishes, trimmed and cut in half, plus thinly sliced
    radish for garnish
4 cloves garlic, peeled and smashed
½ cup apple cider vinegar
¼ cup toasted blanched hazelnuts, for garnish

Place the chicken in a bowl. Add the salt, thyme leaves, mustard, and pepper; toss well to coat.

Preheat the oven to 425°F. Place a large ovenproof sauté pan over high heat. Place a wire rack over a rimmed baking sheet next to the stove.

When the sauté pan is hot, add the oil. Carefully place the chicken in the hot oil in a single layer and cook, flipping once, until well browned on both sides. (You may have to work in batches.) As the chicken is browned, transfer it to the rack to rest.

When all the chicken has browned, carefully pour any excess oil out of the pan and discard.

Place the pan back over medium heat and add 2 tablespoons of the butter. Immediately add the turnips, apple, radishes, garlic, and thyme branches. Toss well and sauté for about 3 minutes, until just softened. Return the chicken to the pan along with any juices that may have collected underneath. Toss everything well to combine. Transfer the pan to the oven.

Roast for about 20 minutes, until the chicken registers an internal temperature of 160°F on an instant-read thermometer. Remove the pan from the oven and transfer the chicken pieces to the wire rack to rest; after resting, the temperature should reach a minimum of 165°F. Add the vinegar and the remaining 1 tablespoon butter to the sauté pan and stir well over low heat to deglaze. Boil 1 minute to thicken slightly.

Divide the chicken among four plates and arrange the vegetables and apple around it. Pour over the sauce. Garnish with the hazelnuts and radish slices.

**SERVES 4**

**Main Subtypes:**

Bell pepper, chile

**Best Pairings:**

Stone fruit, aromatic spices, citrus, mint, wine, cheese, sour cream, yogurt

**Surprise Pairings:**

Pumpkin, peach, dill, cinnamon, rhubarb

**Substitutes:**

Most peppers may be substituted for one another, depending on heat level

Capsicum are members of the nightshade family, which also includes tomato, eggplant, and potato—so it's not surprising that these three ingredients have a particular affinity for one another. The flavor of capsicums is mainly derived from fruity and floral aromas, but the flavors of particular varieties can range from highly vegetal and green (green bell peppers) to sweet and berry-like (aji panca from Peru, whose flavor is often overshadowed by intense heat). The spiciness of chiles is derived from capsaicin, a flavorless compound that produces a burning sensation wherever it comes in contact with the body's soft tissues. Capsaicin is hydrophobic, meaning it cannot be rinsed away by water. But it will bond with fat—which is why drinking milk or even downing a shot of blue cheese dressing is the best way to eliminate the burn of spicy foods.

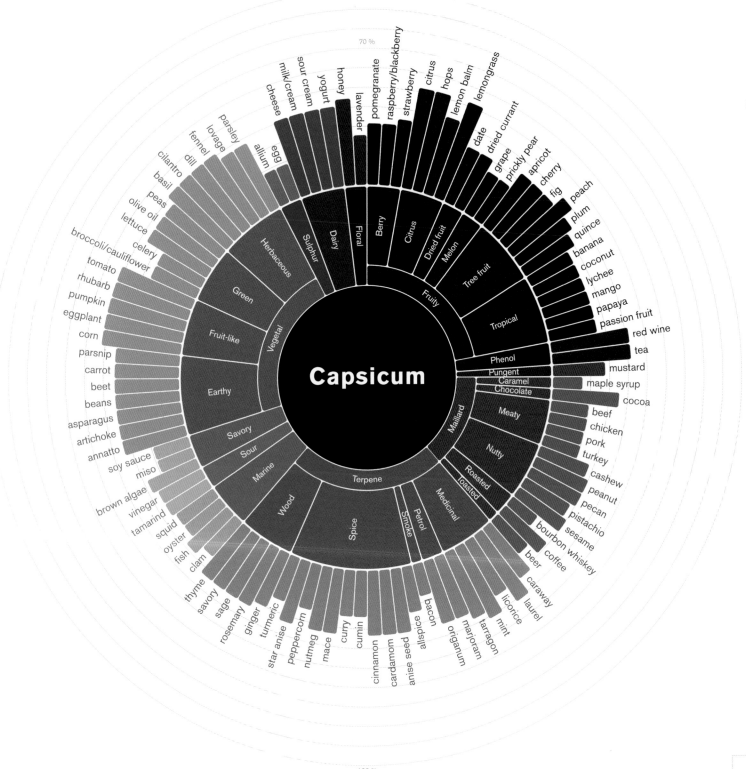

# Banana and
# Chile Hot Sauce

This recipe makes the most of the fruity aromas found in capsicums—and it does so in a most unexpected way. The tropical fruit aromas of banana pair perfectly with chiles and lend a pleasant sweetness to your new favorite hot sauce. Xanthan gum will help thicken and emulsify the sauce, but it's all right to leave it out. Use this as you would your favorite sauce, though it is particularly great with chicken, fish, and shrimp.

1 teaspoon vegetable oil

4 red jalapeño or habanero chiles, minced (for a milder sauce, remove the seeds before mincing)

2 tablespoons minced cilantro stems

4 cloves garlic, minced

½ cup rice vinegar

1 large green (underripe) banana, peeled and chopped (about 1 cup)

⅛ teaspoon xanthan gum (optional)

Heat the oil in small sauté pan over medium-high heat. Add the chiles, cilantro, and garlic and cook until fragrant and tender, but not browned. Add the vinegar and scrape the pan to loosen any stuck-on bits. Mix in the banana and continue cooking until the vinegar is reduced by half, about 1 minute.

Transfer to a blender or food processor and purée until smooth. With the machine running, add the xanthan gum (if using) and blend for 10 seconds more to thicken and emulsify the sauce. Transfer to a jar, seal tightly, and store in the refrigerator for up to 4 weeks. If you omitted the xanthan gum, shake well before using.

**MAKES ABOUT 1 CUP**

Similar to sugar syrup (see page 230), the flavor of caramel is derived largely from Maillard reactions. Toasted, roasted, and nutty aromas are created as sugars and proteins break down under the stress of heat. The longer sugar is left exposed to high temperatures, the darker and more flavorful it will become. However, Maillard reactions have their limits; with prolonged exposure to heat, sugars will burn, creating bitter, acrid flavors. Unlike sugar syrup, caramel has very little in the way of vegetal and berry aromas. Caramel flavor is derived almost exclusively from Maillard aromas.

**Best Pairings:**

Citrus, vinegar, wine, roasted meat, bacon, chocolate, nuts

**Surprise Pairings:**

Fish sauce, tamarind, miso

**Substitutes:**

Molasses, sorghum syrup, maple syrup, coffee, bourbon

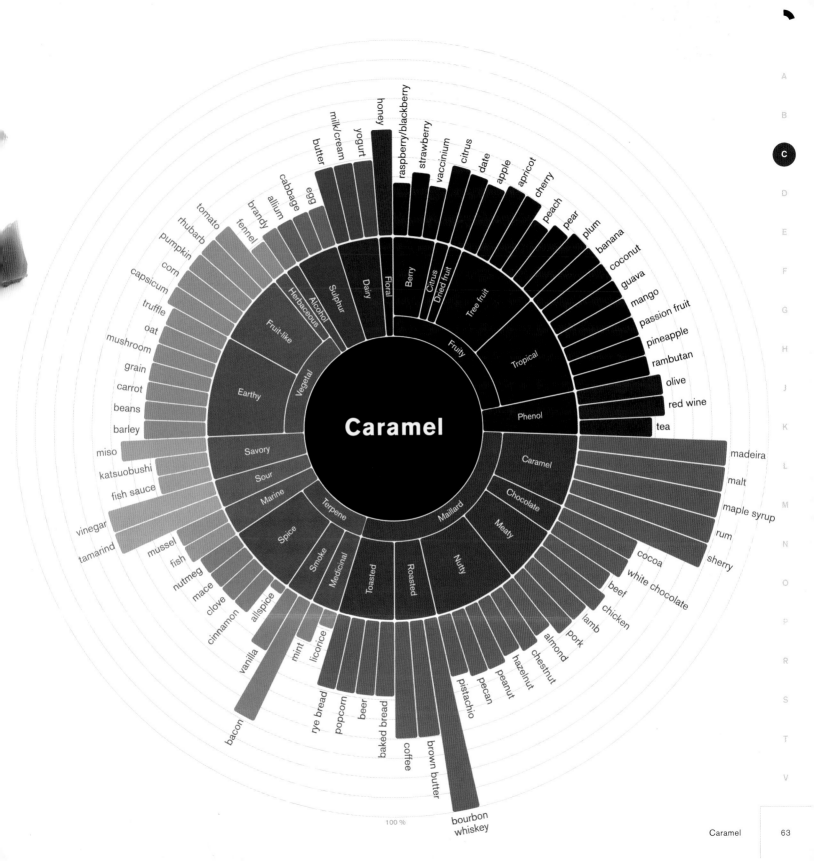

**Caramel**

# Spicy Fish Sauce Peanut Brittle

Brooke and I love this brittle as a topping for steamed buns filled with tender pork or beef. Try it as a savory-sweet, crunchy finish for meat or vegetables, too.

2 cups sugar

6 tablespoons fish sauce

4 dried arbol chiles or 2 teaspoons red pepper flakes

1 cup coarsely chopped peanuts

Line a baking sheet with wax paper and grease the paper with nonstick cooking spray.

In a heavy saucepot, stir together the sugar, fish sauce, and chiles until the sugar is completely moistened. Place the pot over medium heat and cook, swirling the pot occasionally as the sugar melts; do not stir. If you notice crystals forming around the side of the pot, wipe the inside with a brush moistened with water to wash the crystals back into the caramel.

Continue cooking the caramel at a simmer until it has a deep brown color, 12 to 15 minutes. Carefully judge the color as the fish sauce will make the caramel look darker than it really is. The temperature should register between 330° and 350°F on a candy or deep-fry thermometer. As soon as the caramel is done, remove the pot from the heat and stir in the peanuts.

Immediately pour the mixture out onto the wax paper. Be careful, as it will be extremely hot. Carefully tilt the pan to spread the caramel into a thin layer, or spread with a heat-proof spatula. Let cool completely to harden, then break into small portions. Store in an airtight container.

MAKES ABOUT 2 CUPS

**Best Pairings:**

Citrus, wine, vinegar, radish, olive, cilantro and coriander seed, cardamom, star anise, coffee, cilantro, basil, mint, tarragon

**Surprise Pairings:**

Balsamic, basil, coffee, eucalyptus

**Substitutes:**

Radish, parsnip, celery, sweet potato

Wild carrots were first cultivated more than 5, ago, almost exclusively for their leaves and se because the plant produced a bitter, woody ro modern carrot that we eat today—a sweet, te vegetable—is the result of centuries of select ing. The orange carrot we know is also some newcomer; the original carrot varieties were yellow. Woody and resinous aromas still play a role in the flavor of carrot, along with earthy a notes. Although the root is delicious, leafy ca are good, too; they have a mild herbaceous fl a cross between carrot and parsley.

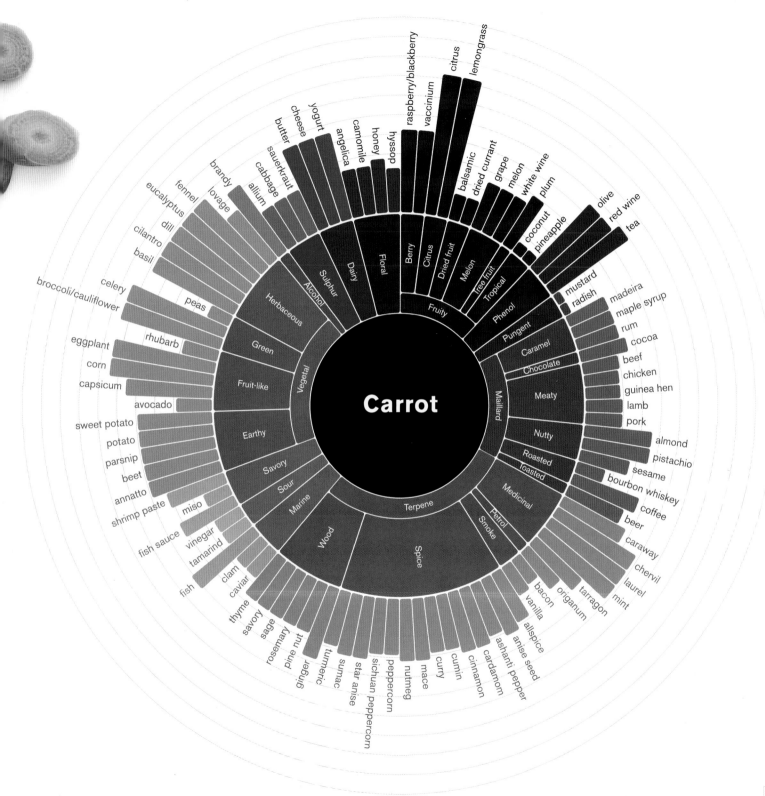

**Carrot**

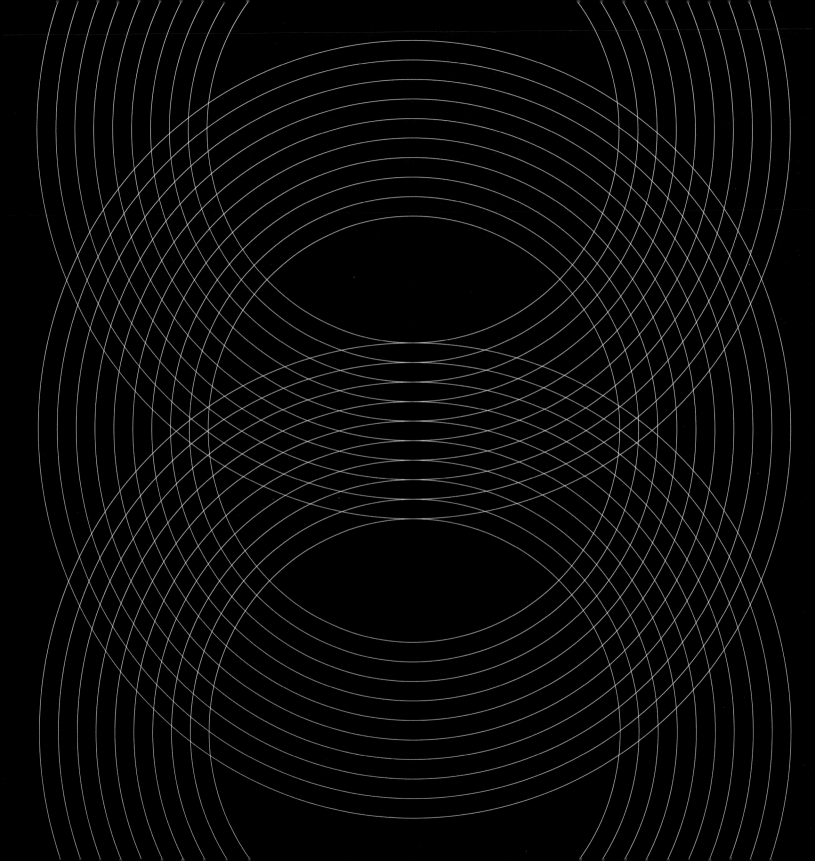

# Coffee and Five-Spice Roasted Rainbow Carrot Salad

Carrots love spice. The aromas that come together to make the flavor of carrot are heavy on spice notes, as well as whiffs of citrus. In this recipe, exotic spices like star anise and cinnamon get an extra kick from coffee, which we use here as a spice. Herbs and yogurt help smooth out the whole dish, and tie together all of these wildly unexpected flavors.

1 pound rainbow carrots, peeled and trimmed
3 tablespoons extra-virgin olive oil, plus more for the arugula
1 teaspoon kosher salt
1 teaspoon ground coffee
½ teaspoon five-spice powder
¼ cup fresh lemon juice, plus more for the arugula
1 tablespoon minced peeled fresh ginger
¾ cup olive oil
4 cups loosely packed arugula leaves
Herbed Yogurt Dressing (recipe follows)
Toasted sliced or slivered almonds, for garnish (optional)

Line a baking sheet with aluminum foil. Place it in the center of the oven and preheat the oven to 425°F.

Combine the carrots, the 3 tablespoons extra-virgin olive oil, the salt, coffee, and five-spice powder in a large bowl and toss to coat the carrots thoroughly. Spread in a single layer on the hot baking sheet, being sure not to crowd them. Roast until the carrots are tender and browned around the edges, about 25 minutes. Remove from the oven and cool slightly on the baking sheet.

Whisk together the lemon juice, ginger, and the ¾ cup olive oil in a large bowl. Add the carrots and toss to coat.

Lightly dress the arugula with a little extra-virgin olive oil and lemon juice. Toss to coat thoroughly.

Spoon the dressing onto a serving platter. Arrange the carrots on top, and then the arugula so the carrots are still visible under the greens. Top with almonds (if using) and serve immediately.

## HERBED YOGURT DRESSING

2 teaspoons minced chives
2 teaspoons minced fresh dill leaves
Grated zest of 1 lemon
Pinch of kosher salt
Pinch of freshly ground black pepper
1 cup plain Greek yogurt

Combine all the ingredients in a bowl and mix well. Cover and refrigerate until you're ready to serve.

**Makes 1 cup**

**SERVES 4**

**Main Subtypes:**

Orange, lemon, lime, grapefruit

**Best Pairings:**

Cilantro, ginger, bell pepper, cauliflower, broccoli, Brussels sprouts, dark chocolate

**Surprise Pairings:**

Sage, caraway, peanut, pecan

**Substitutes:**

Coriander seeds, cardamom, camomile, apricot, lemongrass, lemon balm

Most varieties of citrus known today are the result of centuries of careful breeding or hybridization. (Mandarin oranges, kumquats, pomelos, Australian limes, and citrons are the common species of non-hybridized citrus still commercially available today.) Citrus flavor is based largely on woody or pine aromas; indeed, the word "citron" is derived from the Greek word for cedar, so similar were their aromas found to be. Citrus flavor is most prominent in the essential oils found in the outer colored layer of the rind known as zest. Citrus pulp is less aromatic but contains the sugars and acid that give the fruit its sweet or sour taste.

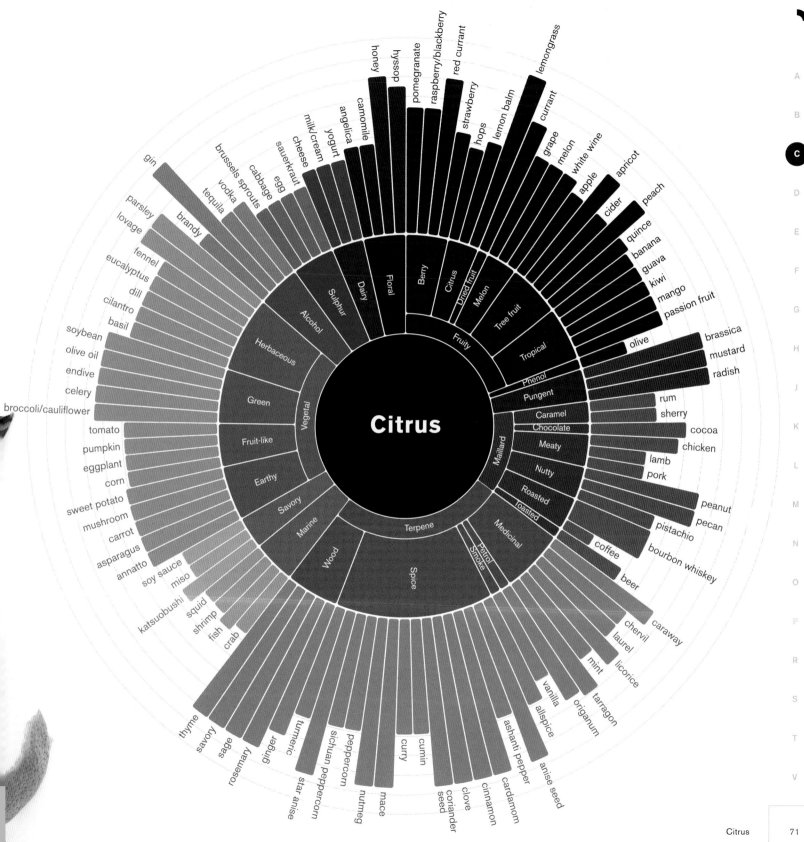

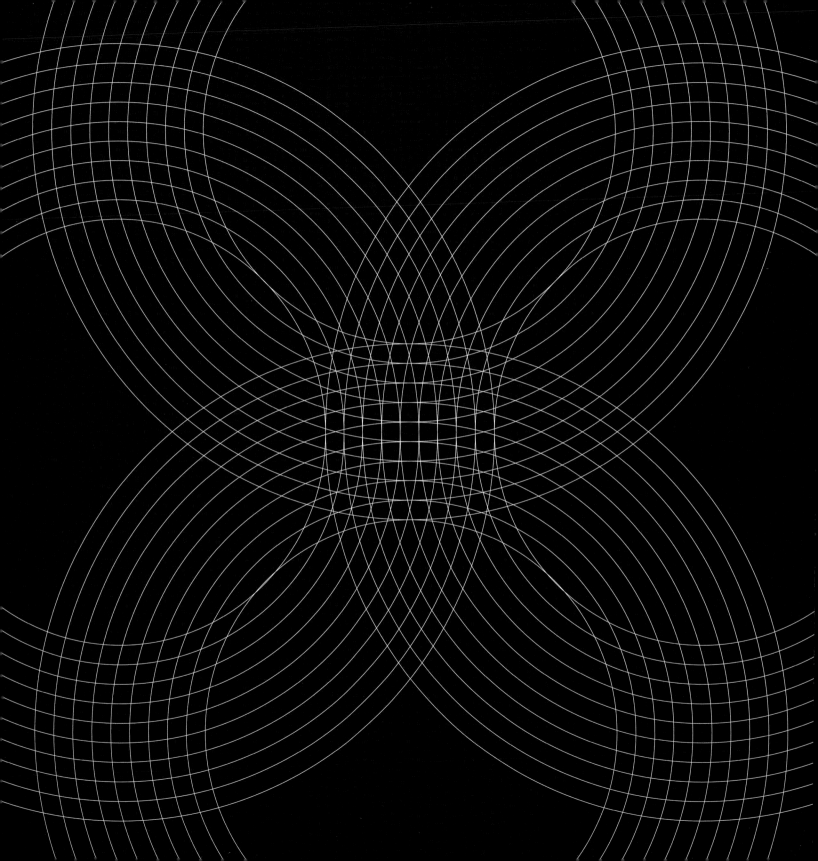

# Preserved Citrus
# with Sage and Chile

Preservation in salt and sugar captures the intense flavor found in the rinds of citrus fruit. Over time, the pulp breaks down and leaves the preserved rinds floating in a pleasantly pungent brine. Preserved citrus rinds may be used to add a punch of flavor to meat and seafood dishes, salads, or cocktails.

2 mandarin oranges

2 lemons

2 limes

2 teaspoons coriander seeds

1 teaspoon fennel seeds

4 branches fresh sage

3 whole dried chiles (such as arbol or Thai) or
    1 teaspoon red pepper flakes

5 tablespoons kosher salt

7 tablespoons sugar

Wash the oranges, lemons, and limes well in warm water. Cut the fruits into quarters and remove any visible seeds. Place the fruits in a bowl and add the coriander and fennel seeds, sage, chiles, salt, and sugar. Mix thoroughly. Pack the mixture into clean glass jars or heavy-duty zip-top bags. If using jars, fill to the top; if using bags, press out all of the excess air. Seal tightly.

Store at cool room temperature (below 70°F) or in the refrigerator to cure for about 3 weeks; refrigerated preserves will take longer to be ready. Occasionally turn the jars or bags to shift the contents. Rinse the citrus rind well before using. The pulp may be used as well, though do so cautiously, as it can be very salty.

**MAKES 1 QUART**

Cocoa beans grow in pods on cacao trees, which are native to South America. The beans have a thin, fleshy, fruit-like covering that has a mild tropical fruit flavor and may be eaten fresh at harvest. The beans themselves are bitter and unpalatable when fresh, however, and must be dried and fermented for five to fourteen days before their distinctive flavor—the result of some 600 aromatic compounds—begins to develop. Not until the Maillard reactions take place during roasting does the beans' true complexity begin to emerge. And when cocoa is turned into chocolate, its flavor can vary further depending on its composition at manufacture: Dark chocolate exhibits typical cocoa characteristics, such as dark-roasted aromas, bitter flavor, and tannins. The addition of dairy to milk chocolate contributes lactones, which have fruity aromas and milder flavor. White chocolate contains only cocoa butter, the lighter, fatty content of cocoa beans. This substance is also odorless, which means that white chocolate's flavor is derived almost entirely from milk and vanilla, rather than from the cocoa beans themselves.

**Best Pairings:**

Coffee, caramel, nuts, whiskey, mushroom, cauliflower, red wine, dried fruit

**Surprise Pairings:**

Beet, avocado, cauliflower, asparagus, fish

**Substitutes:**

Different forms of cocoa/chocolate, toasted nut purées, vegetable ash

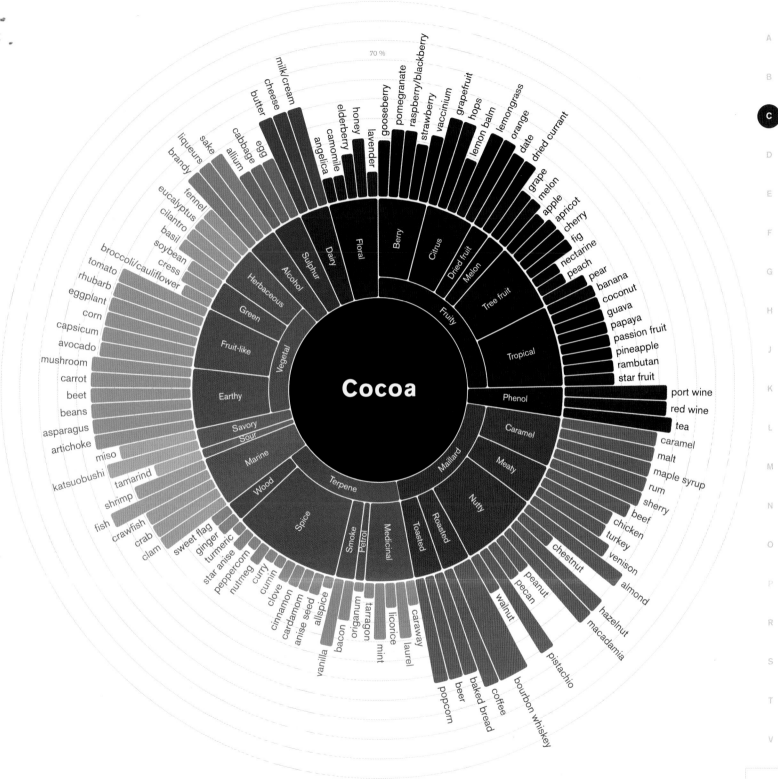

Cocoa

# Chocolate Mousse
# with Crisp Beet Meringue

This recipe delivers unexpected flavors atop a familiar foundation. There is nothing more comforting than a creamy bowl of chocolate mousse. Beets, camomile, and orange zest punch up the flavor of that classic dish, while the beet meringue adds crunch and makes for a dramatic presentation.

## BEET MERINGUE

3 large egg whites
Pinch of cream of tartar (optional)
⅓ cup granulated sugar
¾ cup powdered sugar
¾ cup roasted beet purée (or 1 cup peeled and chopped beets, boiled until very soft, peeled, and puréed in a food processor or blender until smooth)

## CHOCOLATE MOUSSE

¾ cup heavy cream
Grated zest of 1 orange (about 1½ tablespoons)
1 tablespoon dried camomile flowers (optional)
8 ounce semisweet chocolate (at least 70% cacao), chopped
1 tablespoon unsalted butter, softened
6 large egg whites
2 tablespoons granulated sugar

*Make the beet meringue:* Preheat the oven to 150 to 200°F. Line a 13 by 9-inch baking pan with a silicone baking mat or parchment paper and coat with nonstick cooking spray.

Whip the egg whites and cream of tartar (if using) with an electric mixer until frothy. Gradually add the granulated sugar and continue whipping until stiff peaks form and the whites are smooth and glossy. Sift the powdered sugar over the meringue, then gently fold it in.

Place dollops of meringue all over the baking pan. Measure out ½ cup of the beet purée; set aside the remaining ¼ cup for the mousse. Drop spoonfuls of the beet purée in between dollops of meringue, then gently swirl with a spatula and smooth into a thin layer. Bake for about 6 hours at 150°F or 3 hours at 200°F, until crisp but not browned. Remove from the oven and let cool to room temperature. Break into pieces and store in an airtight container in a cool, dry place.

*Make the mousse:* Combine the cream, orange zest, and camomile (if using) in a small saucepot and bring to a boil over medium heat. Turn off the heat, cover the pot, and let steep for 10 minutes.

Combine the chocolate and butter in a heatproof bowl. Strain the cream through a fine-mesh sieve into the chocolate and let stand for 3 minutes. Whisk until all the chocolate has melted and the mixture is smooth. Stir in the reserved beet purée.

Using an electric mixer, whip the egg whites to soft peaks. Gradually add the granulated sugar and continue whipping until stiff peaks form. Gently fold the whipped egg whites into the chocolate mixture one-third at a time until fully incorporated. Refrigerate until well chilled.

When you're ready to serve, divide the mousse among six bowls and top each with pieces of beet meringue.

**SERVES 6**

**Best Pairings:** Lime, avocado, beans, basil, tomato, tropical fruit, mushroom, caramel, nuts

**Surprise Pairings:** Coconut, vanilla, apple, red wine

**Substitutes:** Different forms of corn (grits, polenta, cornmeal), barley, peas

Corn's flavor is the result of a combination of over 200 different aromatic compounds. Sulfur compounds play the largest role, though cooked corn does not present the strong "off" flavors often associated with sulfur. Sweetness and texture also contribute significantly to the overall flavor perception of corn. Corn may be consumed fresh, or dried and ground to make grits (coarsely ground dried corn), polenta (medium ground), or cornmeal (finely ground). Nixtamalization is the process of soaking and cooking corn in an alkaline solution to improve its flavor and nutrition. Nixtamalized corn is known as hominy; it can also be ground and made into masa dough for tortillas or tamales. In this process pyrazines (a group of aromatic compounds associated with roasted flavors) are created. The pyrazines contribute to the unique popcorn-like flavor of nixtamalized corn.

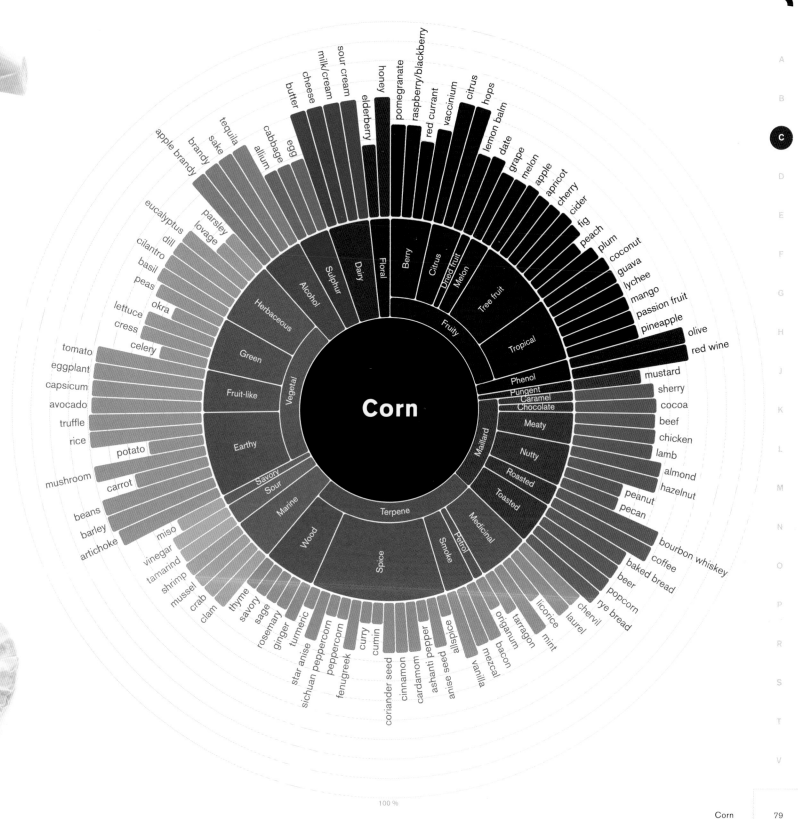

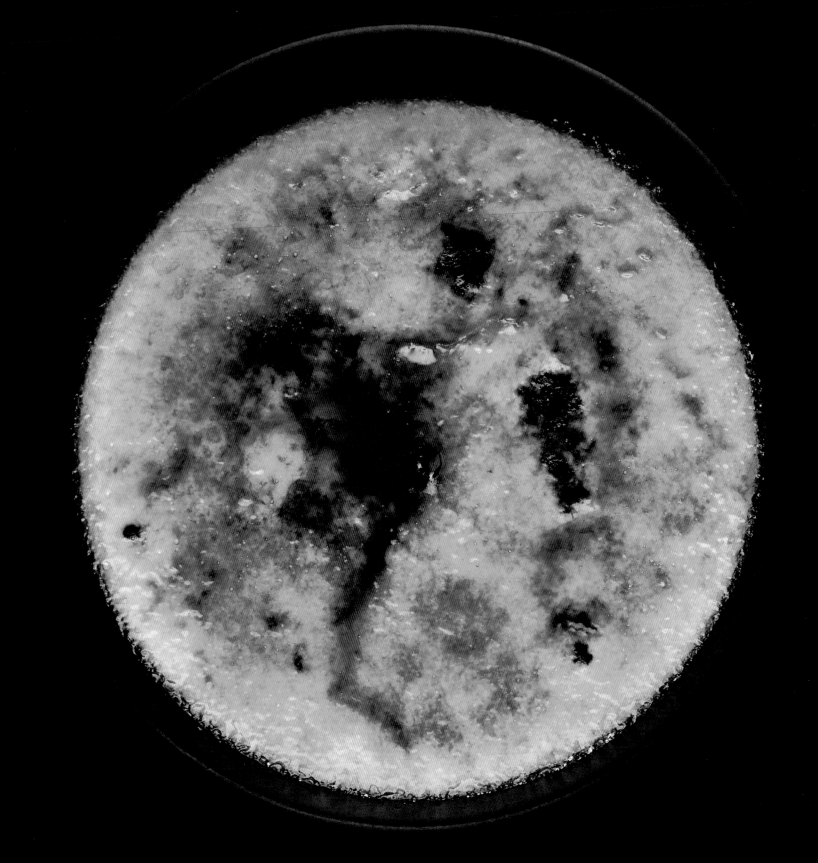

# Corn and Coconut Crème Brûlée

This dessert uses two of corn's more unusual pairings—coconut and vanilla—to make one unforgettable dessert. If you love this combination of flavors as much as we do, try using these ingredients other dishes, such as a cake or cookies.

2 ears fresh corn, husked, silk removed
2 tablespoons unsalted butter
Kosher salt
One 13.5-ounce can unsweetened coconut milk
1 cup water
½ cup granulated sugar
½ cup light brown sugar
2 cups milk
3 large eggs
3 large egg yolks
1 teaspoon pure vanilla extract

Preheat the oven to 325°F. Cut the kernels from the ears of corn. Cut each corncob into 3 pieces.

Melt the butter in a large saucepot over medium heat. Stir in the corn kernels and 1 teaspoon salt. Sauté until just tender, about 5 minutes, being careful not to let the corn or butter brown. Add the corncobs, coconut milk, and water. Bring to a boil, then reduce the heat to a simmer. Cover the pot and cook at a simmer for 15 minutes.

While the mixture simmers, combine the sugars in a blender or food processor, blend well, and set aside.

Remove and discard the corncobs. Transfer the mixture to a blender and blend on high speed until smooth. Strain through a fine-mesh sieve into a large bowl, pressing with a silicone spatula to obtain all the pulp; discard any remaining solids. Whisk in ½ cup of the sugar mixture, the milk, eggs, egg yolks, and vanilla until thoroughly mixed. Adjust the seasoning to taste with salt, if needed.

Divide the mixture among six 8-ounce brûlée dishes or flameproof ramekins, filling them to just below the rim. Place the dishes in a roasting pan or large baking dish. Pour water into the roasting pan to come about halfway up the sides of the dishes. Cover the pan with plastic wrap and carefully transfer it to the center rack of the oven. Bake until just set, about 35 minutes.

Remove the dishes from the pan. Refrigerate until well chilled.

Spread a thin layer of the remaining sugar mixture on top of the baked custards (you may not need it all). Caramelize with a torch. Or place the custards on a baking sheet and caramelize under a preheated broiler. Serve immediately.

**SERVES 6**

Cress refers to a variety of leafy green plants with a distinctive bitter, peppery taste. Included in this group are watercress and garden cress (both genus *Brassica*) and nasturtiums (genus *Tropaeolum*). Cress flavor comes mainly from sulfur compounds, many of which are also found in horseradish and mustard. The sulfur compounds are layered with terpene/wood and vegetal/green aromas.

**Main Subtypes:**

Watercress, nasturtium, garden cress

**Best Pairings:**

Cheese, lemon, orange, berries, grains, bacon, mustard, seafood

**Surprise Pairings:**

Oyster, strawberry, lychee

**Substitutes:**

Arugula, mustard greens, dandelion greens, radicchio

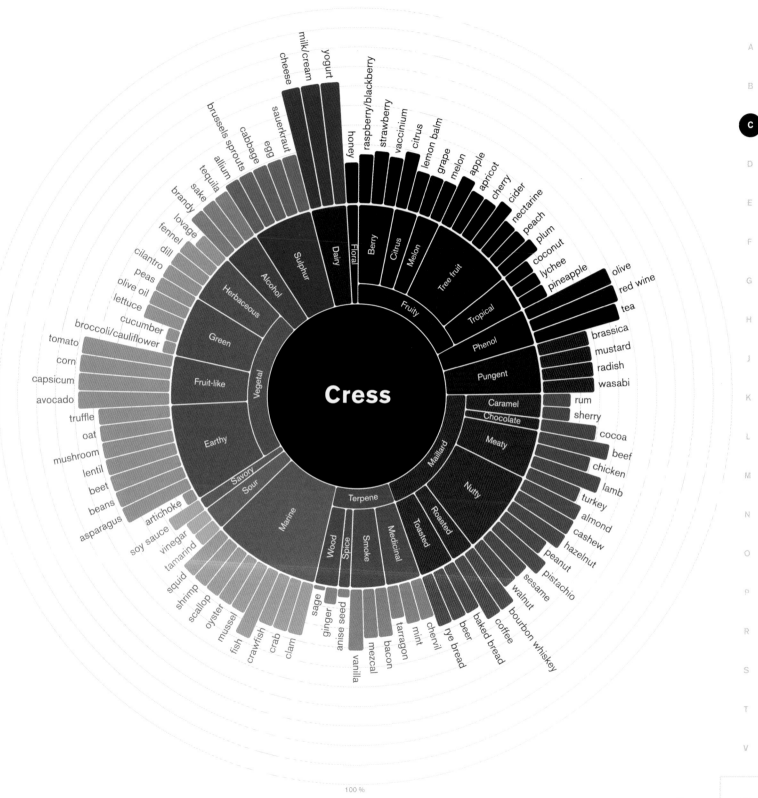

100 %

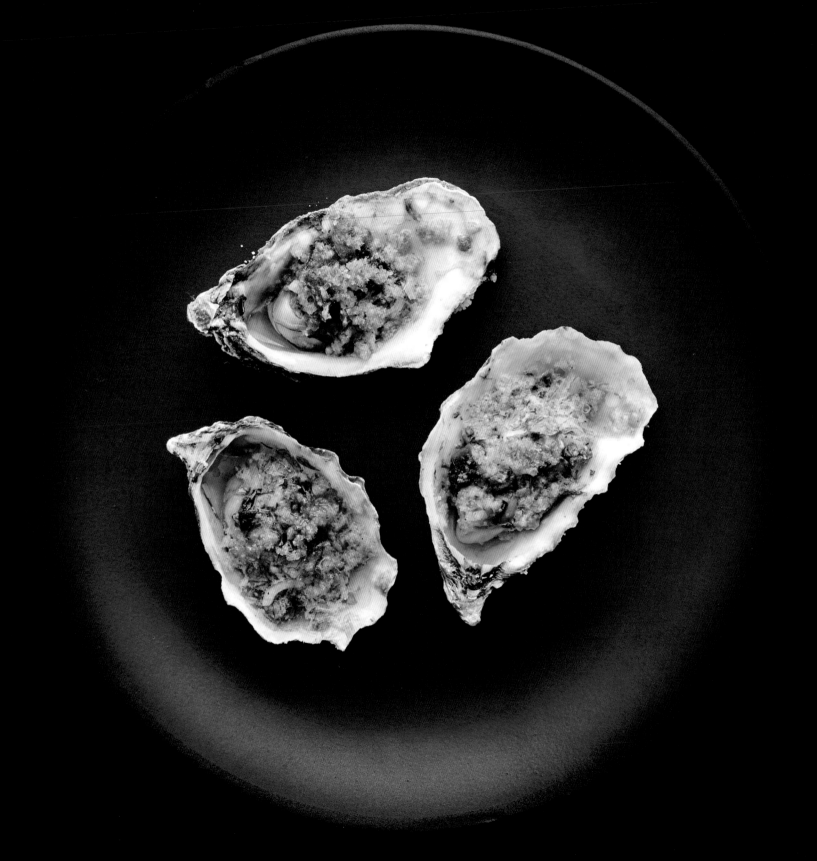

# Grilled Oysters
# with Watercress-Bacon Butter

Watercress-bacon butter is a perfect pairing for oysters, but not only oysters. Make a batch and keep it in the freezer to use on grilled or roasted fish, chicken, and beef. It's also delicious the old-fashioned way: simply spread on bread.

6 strips bacon

2 cups lightly packed watercress leaves

1 tablespoon minced fresh dill

Finely grated zest and juice of 1 lemon

1 shallot, minced

½ pound (2 sticks) unsalted butter, softened

2 teaspoons Dijon mustard

Kosher salt and freshly ground black pepper

Rock salt or kosher salt for grilling

12 oysters, scrubbed clean and opened, top (flat) shells removed

½ cup fresh breadcrumbs

Cook the bacon until crisp. Reserve the fat to add to the butter (if desired) or for another use, or discard. Drain the bacon on paper towels and crumble or chop finely.

Finely chop the watercress by hand and place in a small bowl. Add the dill, lemon zest, and shallot and mix. In a medium bowl, beat together the butter, lemon juice, mustard, and bacon fat (if using) until smooth. Fold in the watercress mixture and bacon. Season to taste with salt and pepper.

Heat a grill until hot. Fill a rimmed baking sheet with a ¼-inch-thick layer of rock salt or kosher salt. Press the oysters into the salt so they sit level. Top each oyster with about 1 tablespoon of the butter and sprinkle with breadcrumbs. Cook on the grill with the lid closed until the butter is melted and bubbling and breadcrumbs are lightly browned, about 8 minutes.

To store any remaining butter, place it on a sheet of plastic wrap or parchment paper and form into a roll 1 to 1½ inches in diameter. Wrap tightly and refrigerate. For longer storage, wrap the roll in aluminum foil and freeze for up to several months. Cut into slices as needed.

**MAKES ABOUT 2 CUPS BUTTER AND 12 OYSTERS**

Crustacean are a group of marine animals that includes over 67,000 species—from culinary classics like crab, shrimp, and lobster to less palatable specimens like wood lice and barnacles. While the cooked meat of crustaceans can vary in texture from animal to animal, the flavor across the class is very similar, with the greatest variations coming from the organisms' environments and diets. The distinctive crustacean flavor is mainly created by nitrogen-containing compounds, specifically amines, which are derivative of ammonia, and thiazines, which also contain a sulfur atom.

**Main Subtypes:**

Crab, lobster, shrimp, crawfish, prawn, langoustine

**Best Pairings:**

Wine, cream, butter, garlic, nuts, breadcrumbs, lemon, apple

**Surprise Pairings:**

Lamb, pumpkin, cherry, annatto seed

**Substitutes:**

Mollusks, fish

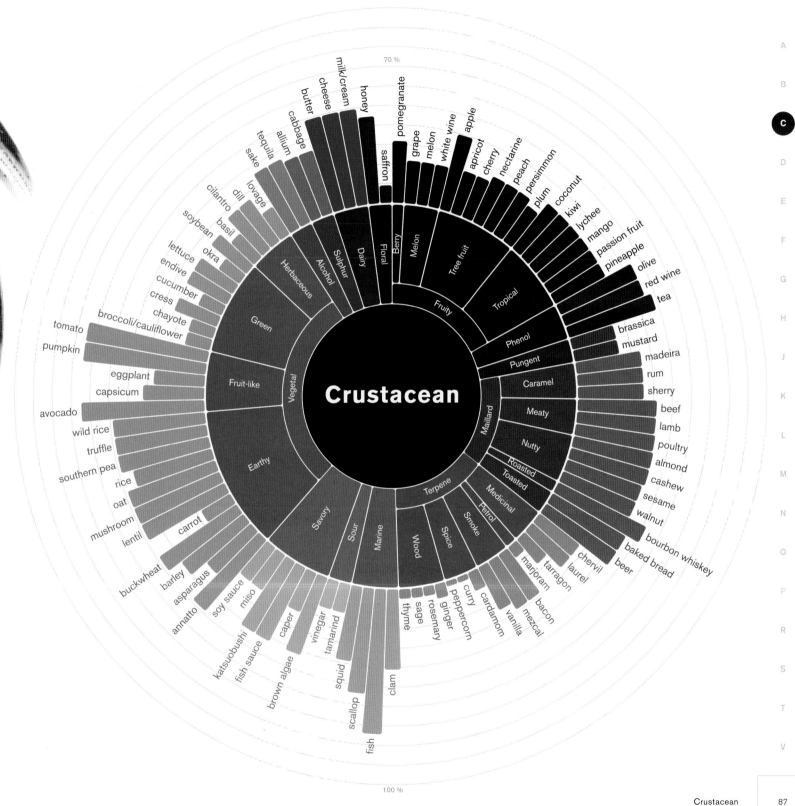

70 %

100 %

**Crustacean**

Fruity
Floral
Dairy
Sulphur
Alcohol
Herbaceous
Green
Vegetal
Fruit-like
Earthy
Savory
Sour
Marine
Wood
Spice
Terpene
Smoke
Petrol
Medicinal
Toasted
Roasted
Nutty
Meaty
Caramel
Maillard
Pungent
Phenol
Tropical
Tree fruit
Melon
Berry

saffron
honey
milk/cream
cheese
butter
cabbage
allium
tequila
sake
lovage
dill
basil
cilantro
soybean
okra
lettuce
endive
cucumber
cress
chayote
broccoli/cauliflower
tomato
pumpkin
eggplant
capsicum
avocado
wild rice
truffle
southern pea
rice
oat
mushroom
lentil
carrot
buckwheat
barley
asparagus
annatto
soy sauce
miso
katsuobushi
fish sauce
caper
brown algae
vinegar
tamarind
squid
scallop
fish
clam
thyme
sage
rosemary
ginger
peppercorn
curry
cardamom
vanilla
mezcal
bacon
marjoram
tarragon
laurel
chervil
beer
baked bread
bourbon whiskey
walnut
sesame
cashew
almond
poultry
lamb
beef
sherry
rum
madeira
mustard
brassica
tea
red wine
olive
pineapple
passion fruit
mango
lychee
kiwi
coconut
plum
persimmon
peach
nectarine
cherry
apricot
apple
white wine
melon
grape
pomegranate

pomegranate
grape
melon
white wine
apple
apricot
cherry
nectarine
peach
persimmon
plum

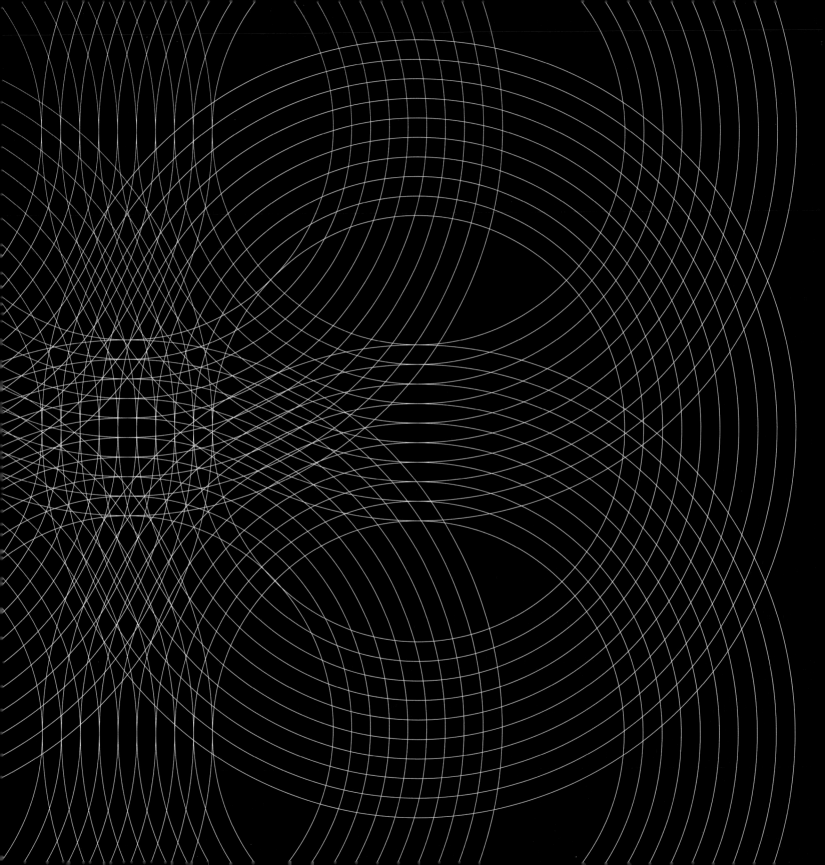

# Shrimp and Lamb Gumbo

Shrimp and lamb is one of the most surprising pairings I have ever come across. I originally discovered this combination when creating a Creole-inspired chickpea-dough dumpling with the Chef Watson team. When the same two flavors are used in a more conventional Creole dish like this rich gumbo, they're perhaps even more delicious.

2 tablespoons vegetable oil

1 pound ground lamb

1 tablespoon kosher salt

2 teaspoons ground turmeric

2 teaspoons filé powder

1½ teaspoons freshly ground black pepper

1 cup diced yellow onion (about 1 medium onion)

½ cup minced celery

½ cup diced green bell pepper

½ jalapeño chile, seeded and diced

1 tablespoon minced garlic (about 3 cloves)

1 cup drained canned chopped tomatoes

4 cups chicken stock

Kosher salt and freshly ground black pepper

1 pound okra, trimmed and sliced

1 pound shrimp, peeled, deveined, and chopped

1 tablespoon fresh lemon juice

1 tablespoon chopped fresh parsley

Heat the oil in a large pot over high heat. Add the lamb, salt, turmeric, filé, and pepper and cook, stirring well to break the meat into small pieces. Cook until all the liquid has evaporated and the meat begins to brown, 8 to 12 minutes.

Stir in the onion, celery, bell pepper, and jalapeño and continue cooking until the vegetables are tender and aromatic, about 6 minutes more. Add the garlic and cook for 1 minute. Add the tomatoes and cook until any juices have evaporated.

Add the chicken stock and bring to a boil. Reduce the heat to simmer and season to taste with salt and pepper. Stir in the okra and simmer for 5 minutes. Add the shrimp and cook 4 minutes more, until the shrimp are firm and white all the way through. Stir in the lemon and parsley just before serving.

**SERVES 4**

Cucumbers are members of the gourd family, making them closely related to pumpkins, squash, and melon. Due to their high water content and low concentration of aroma compounds, however, cucumbers have the mildest flavor of this relatively bland category of ingredients. The key aromas in their flavor are vegetal/green and melon. While raw cucumbers are often sought out for their crisp texture, cucumbers' flavor can actually be heightened by cooking them briefly; the best methods are sautéing or stir-frying.

**Best Pairings:**

Citrus, cilantro, mint, dill, mushroom, tomato, cheese, yogurt, seafood

**Surprise Pairings:**

Vaccinium, hazelnut, persimmon

**Substitutes:**

Crisp lettuce, zucchini, green beans

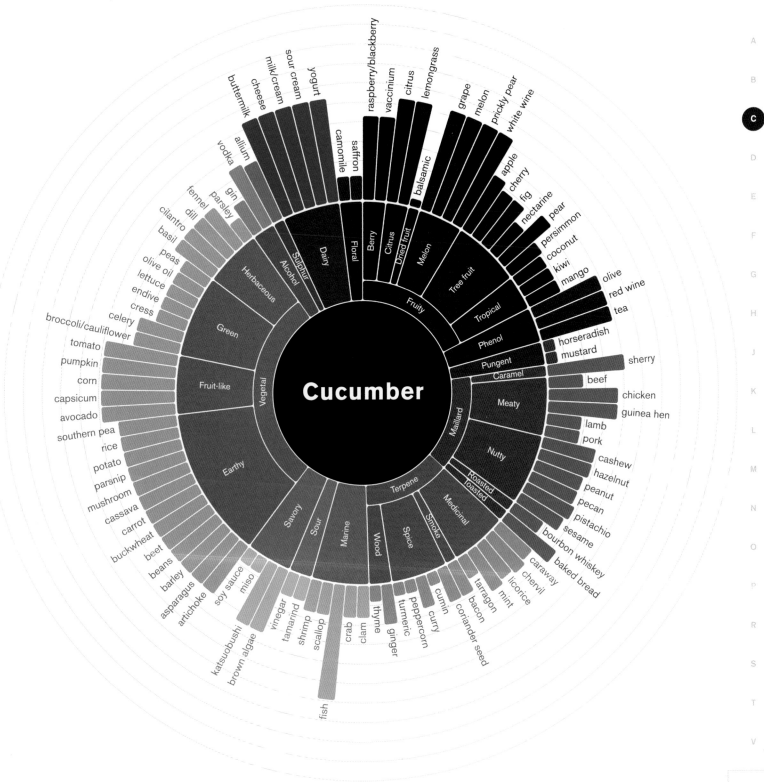

Cucumber

# Cucumber, Berry, and Pistachio Salad with Tamarind Vinaigrette

Cucumbers have a particular affinity for sour flavors, making a tamarind vinaigrette the perfect dressing for this unusual but delicious salad.

2 English (hothouse) cucumbers
½ red jalapeño chile, thinly sliced
½ cup thinly sliced red onion
1 tablespoon chopped fresh dill or mint leaves
¼ cup dried berries (blueberries, cranberries, or a mix), coarsely chopped
Kosher salt and freshly ground black pepper
¼ cup Tamarind Vinaigrette (recipe follows)
¼ cup chopped toasted pistachios

Using a vegetable peeler, peel strips down the length of the cucumbers to create a striped effect. Slice the cucumbers thinly. Combine the cucumbers, jalapeño, onion, dill, and dried berries in a large bowl. Season to taste with salt and pepper and toss well. Cover and store in the refrigerator until you're ready to serve.

Just before serving, dress with the vinaigrette (you will not use it all) and toss well to coat. Divide among four plates and garnish with the nuts.

## TAMARIND VINAIGRETTE

1 clove garlic, minced
1 tablespoon minced peeled fresh ginger
Juice of 1 lime
Fine sea salt
¼ cup tamarind concentrate
½ cup extra-virgin olive oil
½ cup canola oil
Freshly ground black pepper

In a small bowl, whisk together the garlic, ginger, lime juice, and ½ teaspoon salt. Set aside 10 minutes for the flavors to combine.

Whisk in the tamarind. When smooth, whisk in the olive and canola oils until the dressing is emulsified. Season to taste with salt and pepper. Serve immediately, or transfer to glass jar, seal tightly, and store in the refrigerator up to 3 weeks. Shake or whisk to recombine before using.

**Makes about 1½ cups**

**SERVES 4**

**Main Subtypes:**

Milk, cream, cheese, yogurt, buttermilk, sour cream, crème fraîche, butter

**Best Pairings:**

Tropical fruit, roasted meat, dried fruit, honey, toasted nuts, seafood

**Surprise Pairings:**

Date, tamarind, potato chips

**Substitutes:**

Coconut milk, almond milk, kefir, soy milk

The taste of milk and other dairy products depends on a number of factors: the diet of the animal that produced the milk or cream; the production methods; the fat content; and the type of fermentation undergone (if applicable). There is an underlying flavor, however, that is largely consistent across the different products that can be made from milk. This flavor is caused primarily by the sweet and floral aromas of lactic acid esters found in all milk products, as well as sour scents. Soured milk products (yogurt, sour cream, crème fraîche, and buttermilk) get their distinctive tang from higher concentrations of lactic acid, which is mostly odorless— and hence flavorless, except in large quantities. The aromatic compound diacetyl, a byproduct of fermentation, is a secondary factor in the flavor of these products; it has a rich aroma, and is the main component of imitation butter flavoring.

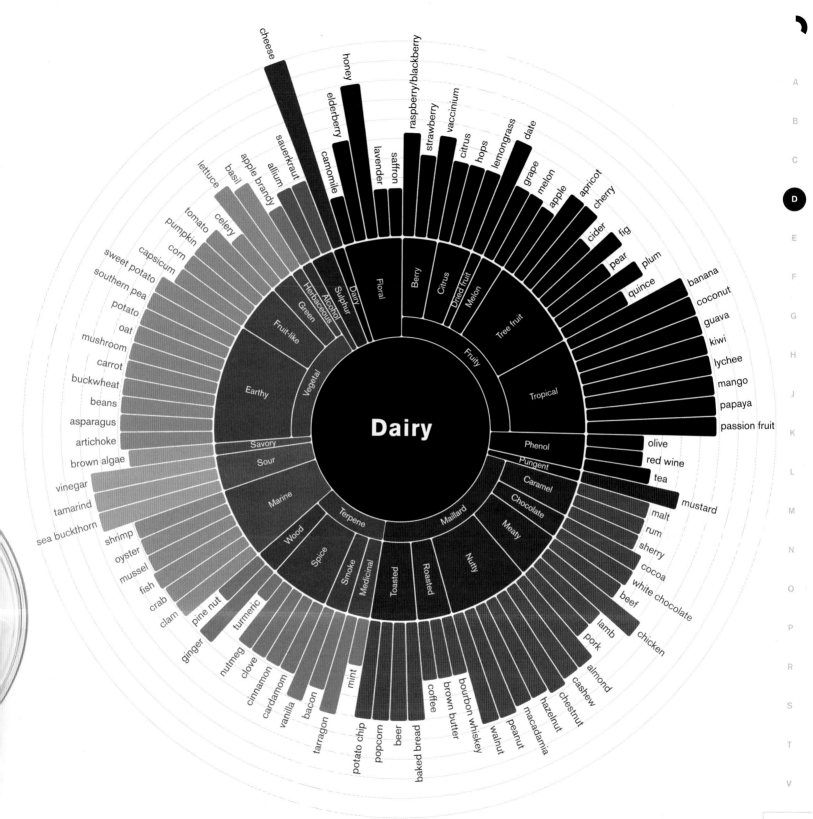

Dairy

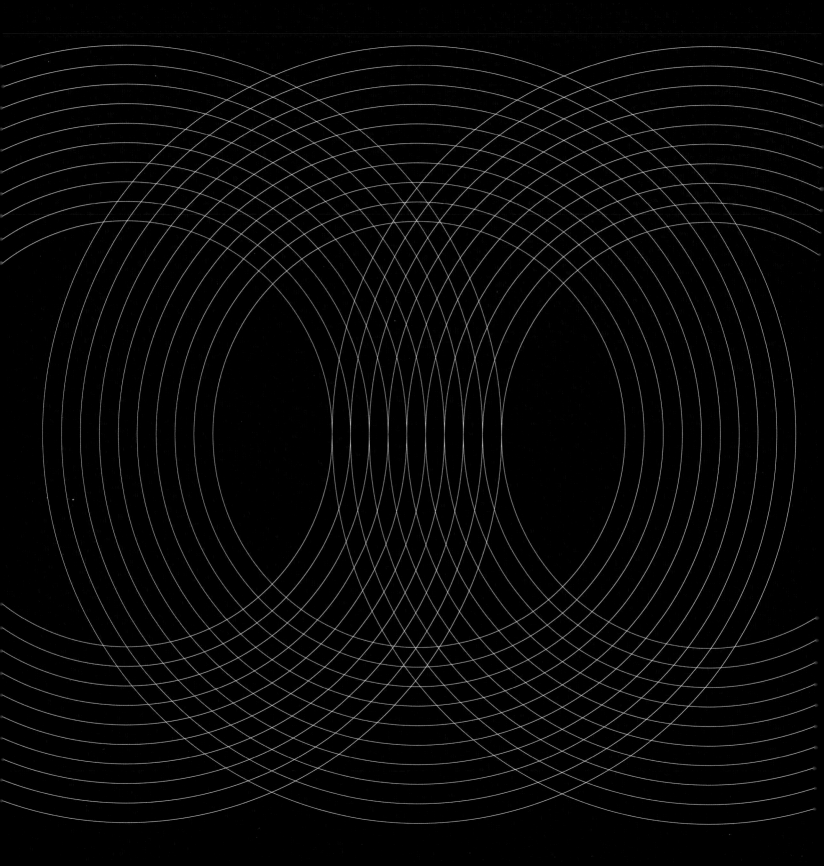

# Walnut and Spiced Yogurt "Hummus"

In this twist on the traditional recipe for hummus, chickpeas are replaced with walnuts and thick Greek yogurt. The result is one of the creamiest and most umami-rich spreads we have ever tasted. Try it as a base under roasted shrimp, as part of a crudités platter, or for a really sexy pita sandwich.

¼ cup extra-virgin olive oil
6 cloves garlic, chopped
2 cups walnut pieces, toasted
Grated zest and juice of 1 lemon
1 teaspoon ground turmeric
Pinch of ground cumin
1 tablespoon white miso (optional)
½ cup plain Greek yogurt
Kosher salt and freshly ground black pepper

Combine the olive oil and garlic in a small sauté pan over medium heat. Cook until the garlic begins to sizzle lightly and smells sweet, 3 to 5 minutes. Remove from the heat before the garlic begins to brown and let cool to room temperature.

Combine the garlic and oil, walnuts, and lemon zest in a food processor and pulse to combine. Add the lemon juice, turmeric, cumin, and miso (if using). Process to a smooth paste. Transfer the paste to a bowl and stir in the yogurt. Season to taste with salt and pepper. Serve immediately or transfer to an airtight container and store in the refrigerator for up to 10 days.

**MAKES 2 CUPS**

**Best Pairings:**

Citrus, cream, cheese, mushroom, truffle, beef, chicken, roasted/smoked meats, seafood, asparagus

**Surprise Pairings:**

Vanilla, carrot, rhubarb

**Substitutes:**

Mayonnaise, vegetable purées, cream, butter; in baking: applesauce

Egg whites are composed of about 90 percent water, 10 percent protein, and trace amounts of fat and nutrients, while egg yolks contain a variety of unsaturated and saturated fatty acids and a range of vitamins and minerals. The hue of the egg yolk is determined by its concentration of lutein, a carotenoid derived from the laying hen's diet. The color of an egg's shell is determined solely by the breed of the animal that produces it. When raw, eggs have little or no flavor, but when they are cooked they release sulfur compounds, which gives them their distinctive flavor. The longer eggs are cooked, the more sulfur compounds they release—which is why hard-boiled eggs (which are usually cooked for too long at a too-high temperature) can have such a strong sulfur odor.

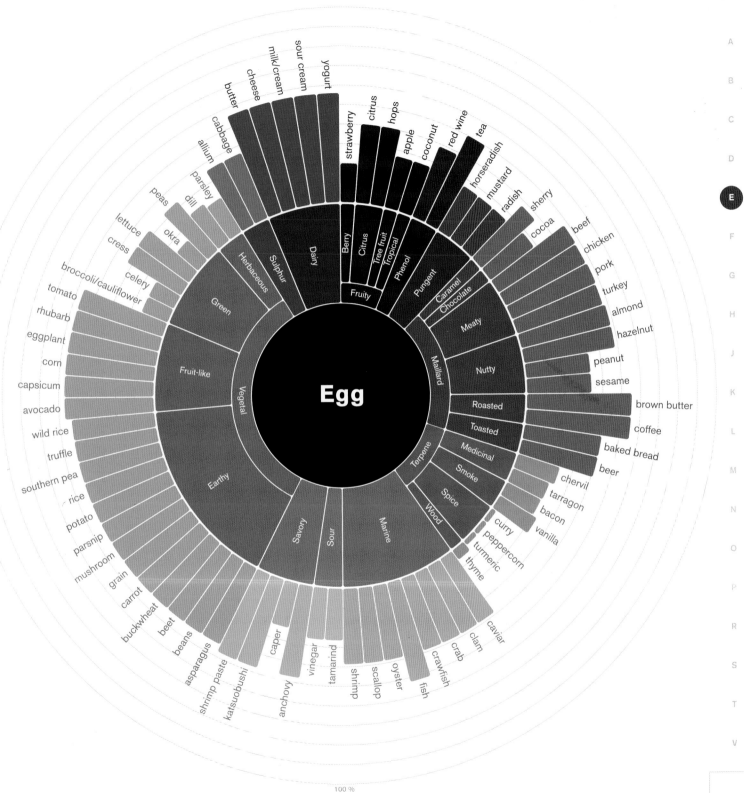

Egg

# Cured Egg Yolks

Cured egg yolks are an excellent way to add egg flavor to a variety of dishes where it would not otherwise be possible. Try grating these cured yolks like cheese over pasta or salad. Or use as a finishing seasoning on cooked meat; porcini-cured yolks (see the Variations) are particularly delicious on grilled beef.

2 cups kosher salt
1 cup sugar
6 large egg yolks

In a small bowl, whisk together the salt and sugar. Spread half of the cure mixture in a shallow plastic container or glass baking dish. Use your thumb to make 6 indentions in the cure. Carefully place an egg yolk in each indention. Use the remaining cure to cover each yolk. Cover the container and refrigerate for 3 days.

Fill a small bowl with room-temperature water, cover a plate with paper towels, and set a wire rack over a baking sheet. One at a time, carefully remove a yolk from the cure and place it in the water to rinse off any remaining cure. Gently transfer to the paper towels and pat dry, then place on the rack. Let the yolks air-dry at room temperature for 2 hours.

Transfer the yolks to a plastic container and seal tightly. Store in the refrigerator for up to 2 weeks.

## VARIATIONS

Citrus: Add 2 tablespoons finely grated citrus zest to the cure mix.

Porcini: Add 1 tablespoon porcini powder to the cure mix.

Spice: Add 1 tablespoon freshly ground black pepper, 1 teaspoon ground coriander, 1 tablespoon fennel seeds, and 1 tablespoon red pepper flakes to the cure mix.

Umami: Add 1 teaspoon porcini powder and ¼ cup crushed bonito flakes to the cure mix.

**MAKES 6 YOLKS**

**Best Pairings:**

Tomato, mushroom, cheese, yogurt, olive,
wine, lemon, orange, broccoli, basil

**Surprise Pairings:**

Walnut, pomegranate, elderberry, melon

**Substitutes:**

Zucchini/summer squash, portobello mushrooms

Early European eggplant cultivars were small, oblong, and white, just like an egg—hence the name. Eggplant is a member of the nightshade family, Solanaceae, which includes tomatoes and potatoes; because eggplants and tomatoes prefer similar growing conditions, they are often found together in recipes. Eggplants have very little aroma and are largely inedible when raw; when cooked, they derive the majority of their flavor from Maillard reactions and the earthy aromas that are common in other cooked vegetables. Contrary to popular belief, salting raw eggplant will not help to make the flesh less bitter. Rather, salting before cooking helps collapse cell walls, reducing the spongy structure of the eggplant and making it less likely to absorb excess liquid during cooking. So when roasting or grilling eggplant, salting ahead of time is unnecessary; salting should be reserved for frying and sautéing eggplant, to lessen oil absorption.

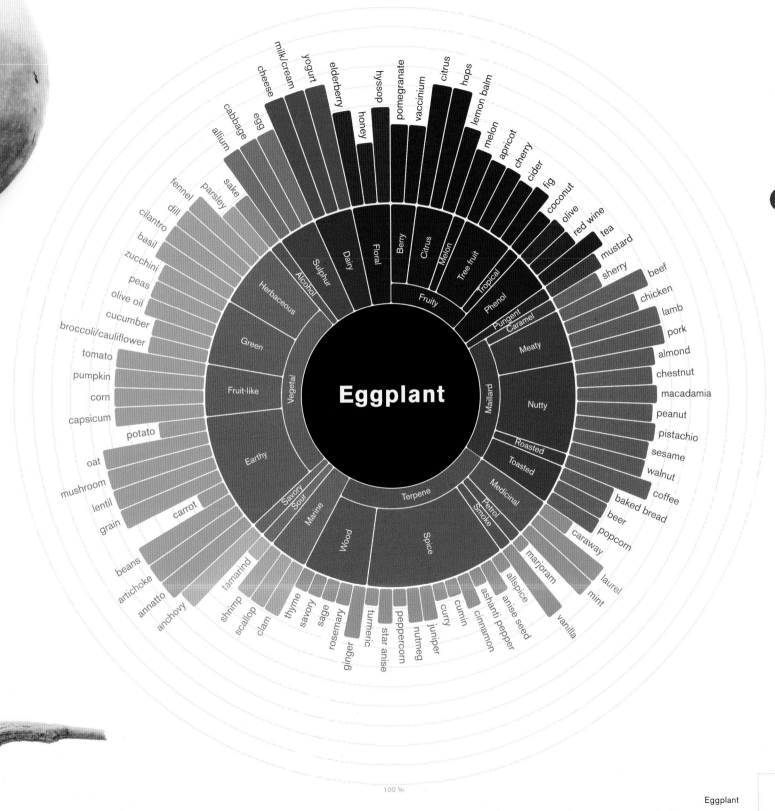

Eggplant

# Fried Eggplant with Muhammara

We found the pairing of eggplant and pomegranate enticing and exotic, although it is not uncommon in Persian cooking. Instead of fresh pomegranate, we deliver their flavor in the form of *muhammara*, a red pepper and walnut sauce augmented with pomegranate syrup. The fried eggplant slices make a great hors d'oeuvre when simply topped with a dollop of muhammara, or go all out and make it like eggplant parm: Alternate layers of eggplant and muhammara in a baking dish, top with a salty sheep's milk cheese—pecorino or feta—and bake.

## FRIED EGGPLANT

2 pounds eggplant (Italian or Japanese), trimmed and sliced ½ inch thick

Kosher salt and freshly ground black pepper

2 cups all-purpose flour

2 large eggs

3 cups dry breadcrumbs

Canola or vegetable oil, for frying

About ½ cup Muhammara (opposite)

¼ cup torn fresh basil leaves, for garnish

Shaved pecorino or Parmesan cheese, for garnish

Lay out the eggplant slices on a paper towel–lined baking sheet and season with salt and pepper. Let rest for at least 10 minutes.

Set up three bowls or small baking dishes. Put the flour in the first bowl. Crack the eggs into the second; add ¼ cup water and a pinch of salt and beat thoroughly. Place the breadcrumbs in the third. Set out a clean baking sheet at the end of the line.

Pat the eggplant slices dry. Working one slice at a time, dip into the flour to coat, then pat off any excess. Dip the slice into the eggs, again coating it; lift out and let the excess drip back into the bowl. Finally, press into the breadcrumbs to coat one side; flip to coat the other side. Set aside on the clean baking sheet while you bread the remaining slices.

Set a wire rack over a baking sheet and place it next to the stove. Fill a wide sauté pan with ½ inch oil and place over medium heat. Heat the oil to 350°F (check with a thermometer), or until a few breadcrumbs sprinkled into the oil sizzle immediately. Add as many eggplant slices as will fit without touching. Cook until golden brown on one side, about 2 minutes, then flip and brown the second side. Transfer to the rack to drain while you fry the remaining slices.

When all the eggplant is cooked and drained, arrange the slices on a platter. Spoon on the muhammara. Garnish with the basil and cheese and serve immediately.

## MUHAMMARA

1 pound red bell peppers

1 cup walnut pieces, toasted

1 tablespoon pomegranate molasses

2 tablespoons extra-virgin olive oil

1 teaspoon ground cumin

Kosher salt

1 teaspoon Aleppo pepper or sriracha, or to taste

Roast the peppers using your favorite method. Cover and let cool to room temperature, then remove the stems, skins, and seeds. Coarsely chop, then add to a blender with the walnuts, pomegranate molasses, olive oil, and cumin. Process until smooth. Season to taste with salt and Aleppo pepper. Serve immediately, or transfer to a glass jar or plastic container, seal tightly, and refrigerate for up to 10 days.

**Makes 2 cups**

**SERVES 4**

Fennel is a member of the botanical family that includes carrots, celery, and parsley. Like carrots and celery, fennel may be consumed raw or cooked; also like carrots and celery, the entire fennel plant has viable culinary uses, from the bulb to the tops and feathery leaves. Fennel seeds, flowers, and even pollen find their way into dishes around the world. The compound anethole gives fennel its distinctive anise- or licorice-like flavor, and limonene—the most prominent compound in citrus—plays a supporting role in fennel flavor.

**Best Pairings:**

Citrus, basil, dill, cilantro, ginger, lemongrass

**Surprise Pairings:**

Vaccinium, sage, rosemary

**Substitutes:**

Celery, lovage, caraway, hoja santa, anise seed

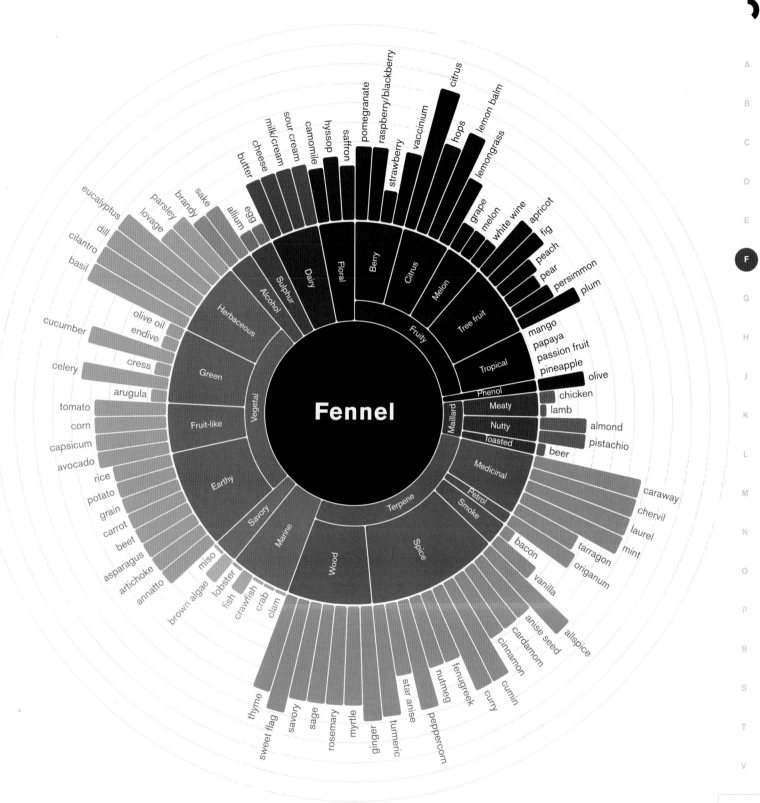

Fennel

# Bourbon-Roasted Fennel and Plums with Coffee Ice Cream

Fennel doesn't often find its way into dessert. But when it's paired with the sweetness of plums, and when bourbon and coffee lend their roasted and toasted aromas to that offbeat combination, this aromatic vegetable works perfectly as a sweet finish to any meal.

1 pint coffee ice cream

2 tablespoons unsalted butter

2 tablespoons light brown sugar

2 plums, cut in half and pitted

1 cup thinly sliced fennel bulb

¼ cup bourbon

½ teaspoon kosher salt

Micro basil leaves, for garnish (optional)

Toasted pecans, for garnish (optional)

Scoop the ice cream into four bowls and place in the freezer until you're ready to serve.

Melt the butter in a sauté pan over medium heat. Cook the butter until the foaming subsides and it takes on a light brown color. Add the sugar and stir until it has melted. Place the plum halves cut side down in the pan and cook for 3 minutes, until lightly browned. Scatter the fennel over the plums and gently toss to coat with the brown sugar mixture. Cook for 3 minutes more, stirring occasionally, until the fennel is tender. Remove from the heat and stir in the bourbon and salt. Transfer to a bowl and let cool for 5 minutes, then spoon over the ice cream. If you like, garnish with basil and/or pecans. Serve.

**SERVES 4**

**Best Pairings:**

Lemon, orange, berries, roasted or smoked meat, wine, tea

**Surprise Pairings:**

Avocado, clam, tomato, capsicum

**Substitutes:**

Apricot, date, cherries, cantaloupe

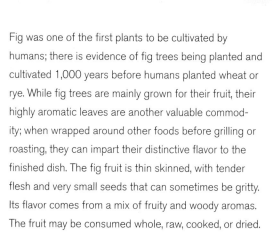

Fig was one of the first plants to be cultivated by humans; there is evidence of fig trees being planted and cultivated 1,000 years before humans planted wheat or rye. While fig trees are mainly grown for their fruit, their highly aromatic leaves are another valuable commodity; when wrapped around other foods before grilling or roasting, they can impart their distinctive flavor to the finished dish. The fig fruit is thin skinned, with tender flesh and very small seeds that can sometimes be gritty. Its flavor comes from a mix of fruity and woody aromas. The fruit may be consumed whole, raw, cooked, or dried.

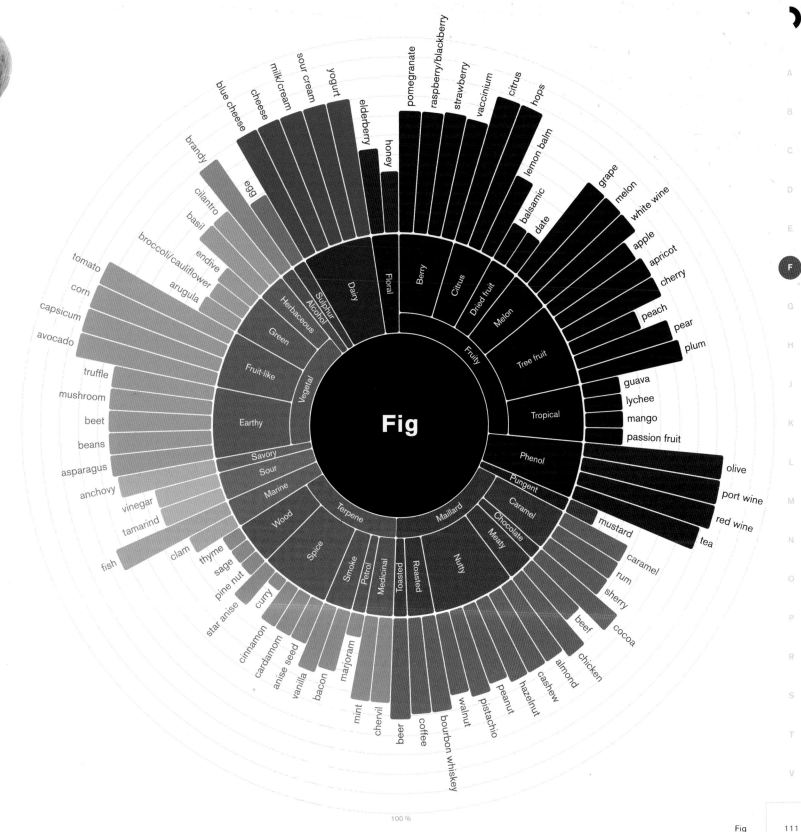

Fig

Fig    111

# Fig, Olive, and Walnut Relish

This relish makes a wonderful accompaniment to cheese and charcuterie. It is also delicious on grilled or roasted chicken or fish, avocado toast, or even a simple cauliflower steak.

½ cup niçoise olives, cut in half and pitted
Grated zest and juice of 1 lemon
½ teaspoon fresh thyme leaves
Pinch of sugar
2 tablespoons extra-virgin olive oil
¼ cup chopped toasted walnuts
1 cup diced fresh Black Mission or Brown Turkey figs
Kosher salt and freshly ground pepper

Combine the olives, lemon zest and juice, thyme, sugar, and olive oil in a bowl. Mix thoroughly. Add the walnuts and figs and gently fold to incorporate them. Season to taste with salt and pepper. Serve immediately, or transfer to a glass jar or plastic container, seal tightly, and store in the refrigerator for up to 10 days.

**MAKES ABOUT 1¾ CUPS**

Fig    113

**Best Pairings:**

Lemon, orange, coconut, melon, toasted
nuts, breadcrumbs, olive, caper, cream

**Surprise Pairings:**

Melon, peanut, coffee

**Substitutes:**

Different varieties of fish may be substituted
for one another.

Like crustaceans, the flavor of fish comes from nitrogen-
and sulfur-based aromatic compounds. Also like crustaceans, different types of fish share many underlying
flavors and thus can be paired according to the same set
of rules. For culinary purposes, however, it can be helpful
to categorize fish in several simple ways. First of all, fish
are either fresh- or saltwater. Fish that live in a saltwater
environment produce more glycine and glutamate
(umami-creating amino acids) to counteract the salinity
in their environment. In general, therefore, saltwater fish
tend to have a richer, fuller taste than their freshwater
cousins. Second, fish can be considered either fatty
or lean. Much of a fish's flavor is carried by the natural
oils in its flesh. Lean fish thus have a milder flavor, while
fatty fish taste stronger and more "fishy." Freshness is
essential in fatty fish, as the natural oils (fatty acids) are
quick to break down and go rancid, which creates off
flavors in the fish. Finally, fish can be classified as high
activity or low activity. Like most animals, the more a fish
uses its muscles, the more flavorful it tends to be. Highly
active fish like mahi mahi, salmon, mackerel, cobia, and
tuna tend to be more flavorful than low-activity fish like
haddock, pollock, and cod. Bass, snapper, grouper, and
tilefish are considered medium-activity fish.

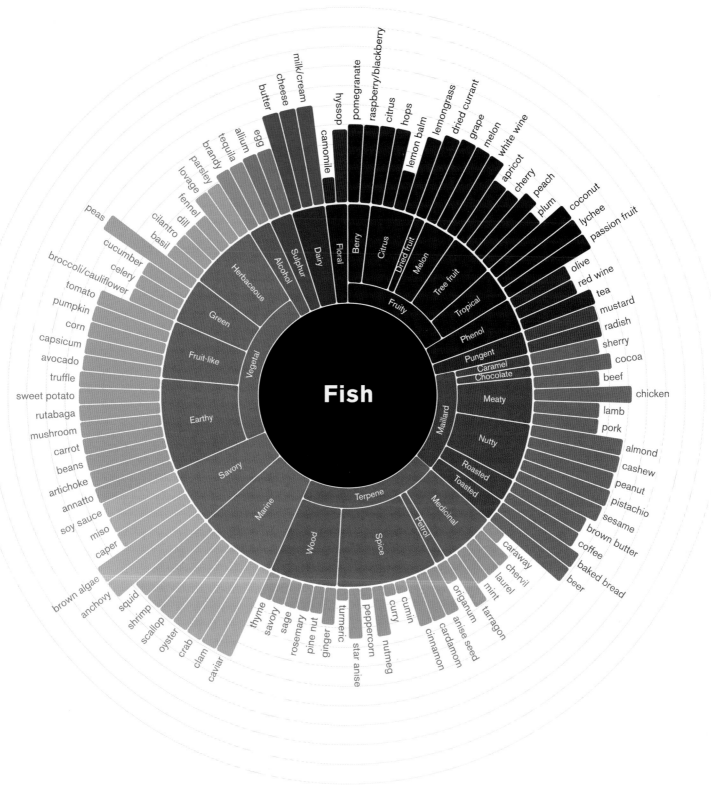

# Fish

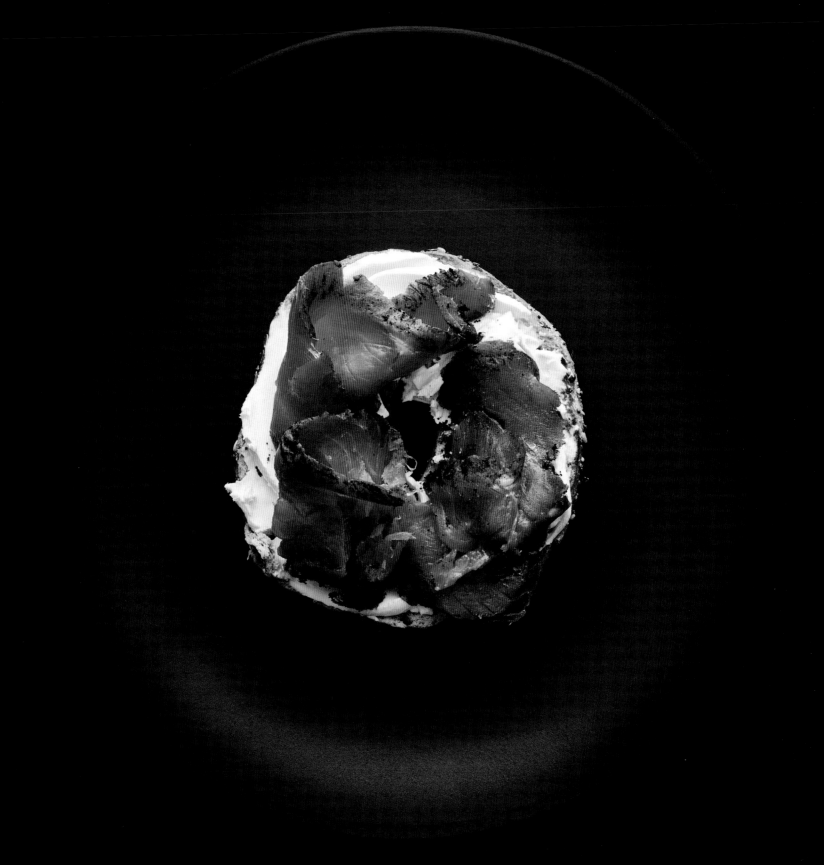

# Coffee-Cured Salmon

This home-cured fish makes for the ultimate weekend bagels-and-lox buffet. After all, who's not already having coffee with their favorite morning fish? Serve on toasted everything bagels with cream cheese, or plate with mint, olives, pomegranate seeds, and toasted sesame seeds to serve as an appetizer.

About 3 tablespoons ground coffee
1 pound skinless salmon fillet
1 cup kosher salt
½ cup sugar
Grated zest of 1 lemon
Olive oil

Rub the coffee liberally onto both sides of the salmon. Set the salmon aside on a plate while you prepare the cure mix.

Combine the salt, sugar, and lemon zest in a small bowl and mix well. Place a piece of plastic wrap about 18 inches long on the work surface. Mound half of the cure mix in the center of the plastic and press it into a rectangle slightly bigger than the piece of salmon. Place the salmon on top and cover with the remaining cure mix. Wrap tightly in the plastic and place in a shallow baking dish. Refrigerate for at least 6 hours or up to overnight.

Unwrap the fish and quickly rinse under cold water. Pat dry with paper towels. Place on a rack set over a pan or plate and refrigerate, uncovered, for 8 to 12 hours to dry. Then wrap tightly in plastic or seal airtight in a plastic bag until ready to serve.

To serve, cut the fish into slices about ¼ inch thick. Lightly brush the slices with olive oil and place them between two sheets of plastic wrap. Pound very lightly until the slices are ⅛ inch thick. Store any leftovers tightly wrapped in the plastic in the refrigerator for up to 2 weeks.

**SERVES 8**

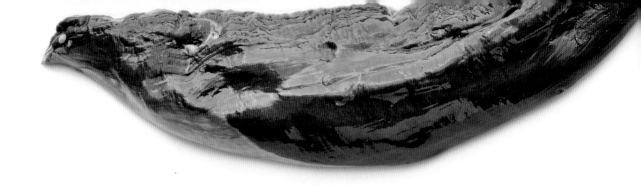

Many animals that historically were hunted for sport and nourishment—and are classified as "game"—have been domesticated much like any other livestock. The flavor of deer, elk, and boar flesh can vary dramatically in the wild because the animals' environments are so variable. Those raised on farms have a milder and more consistent flavor. This trend toward domestication of different game species has led to more commonalities between the meats. Venison, the meat of deer, is a deep red color and extremely lean. It has a robust meaty flavor and relatively mild "game" flavor. Elk, although a separate breed from deer, often has a flavor profile very similar to venison. Boar is generally redder, leaner, and more intensely flavored than pork, and also has a greater concentration of essential amino acids. Wild pigs grow more slowly in captivity than their relative the domesticated pig, and they also tend to yield less meat; however, the meat they produce is thought to be of higher nutritional value. What's more, the compounds responsible for the typical flavors of game meats in the wild are not particularly appetizing on their own; in isolation, these compounds have pungent, fecal, or solvent-like aromas. But they become delectable when mixed with the meat and fat aromas typical of beef and lamb that are prevalent in the flavor of game meat.

**Main subtypes:**

Venison, elk, boar

**Best Pairings:**

Mushroom, parsnip, winter squash, bourbon, coffee, rosemary, juniper, red wine

**Surprise Pairings:**

Eucalyptus, cilantro, coconut, pineapple

**Substitutes:**

Any game meat may be substituted for another. Also: lamb, beef

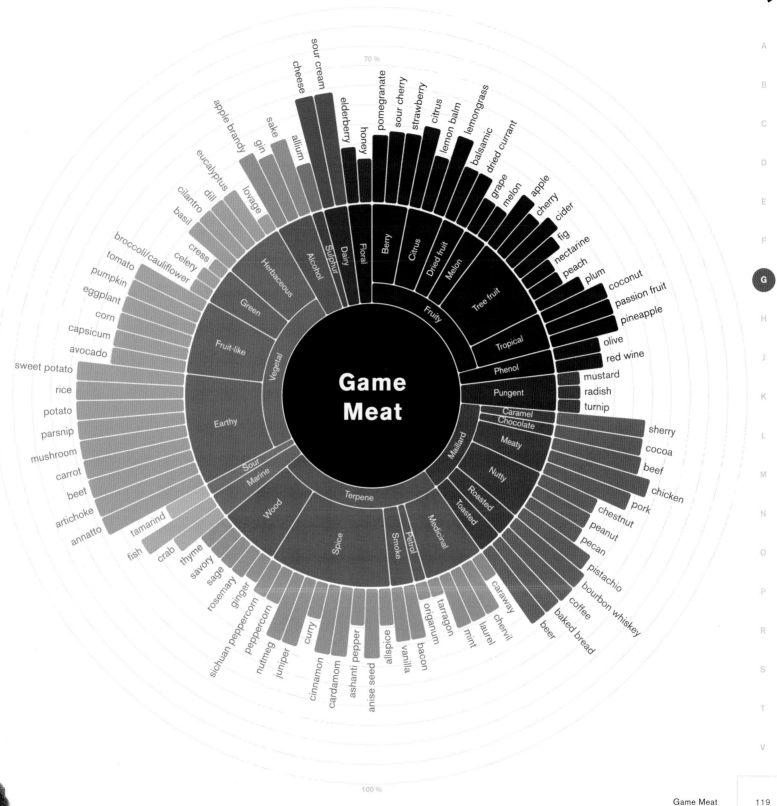

# Game Meat Seasoning

The citrus and spice notes in this rub are the perfect complement for game meats like venison, elk, and boar. Simply rub the mixture generously onto the meat before cooking. It is also delicious on meaty fish like tuna or swordfish.

2 tablespoons black peppercorns
2 tablespoons juniper berries
1 tablespoon coriander seeds
1 tablespoon dried rosemary leaves
2 teaspoons dried grated orange peel
3 tablespoons kosher salt
1 tablespoon sugar

Combine the peppercorns, juniper berries, coriander seeds, rosemary, and orange peel in a mortar and grind with a pestle until thoroughly broken down but not a fine powder. Or grind them together in a spice grinder. Transfer to a small bowl and mix in the salt and sugar. Store in an airtight container in a cool, dark place for up to 3 months.

**MAKES ABOUT ½ CUP**

Ginger belongs to the family Zingiberaceae, which includes other pungent spices like turmeric, cardamom, and galangal. Since plants from the same family tend to pair together very well, it's not surprising that these ingredients often end up together in the same dishes. The knobby root of the ginger plant has a pungent flavor with strong aromas of wood or pine and spice. Gingerol, the compound that gives raw ginger its spiciness, is a relative of capsaicin, the active component in chiles. Cooking transforms gingerol into zingerone, a compound with less spice and a strong pine aroma.

**Best Pairings:**

Alliums, citrus, pine nut, origanum, peppercorn, basil, cilantro, parsley

**Surprising Pairings:**

Cumin, pistachio, pumpkin

**Substitutes:**

Lemongrass, rosemary, galangal, Sichuan peppercorns

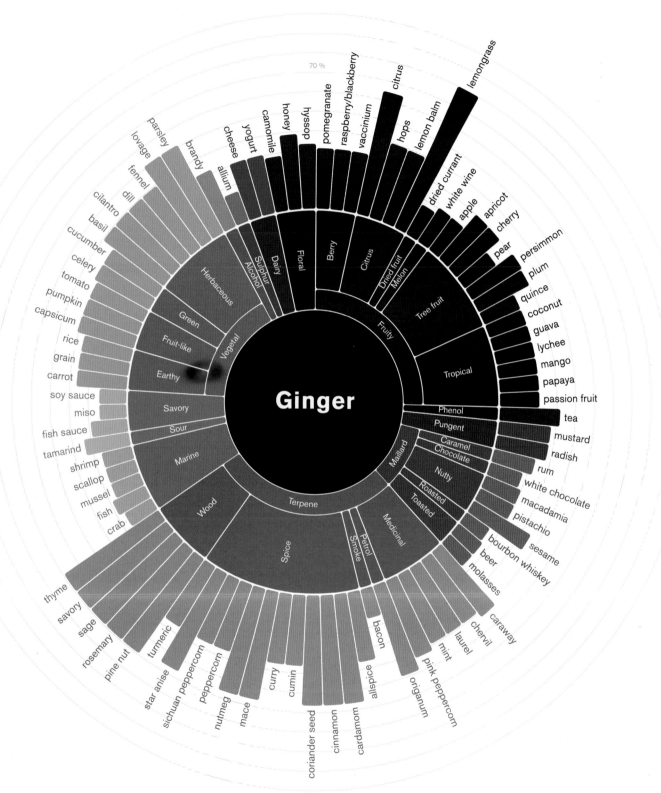

Ginger

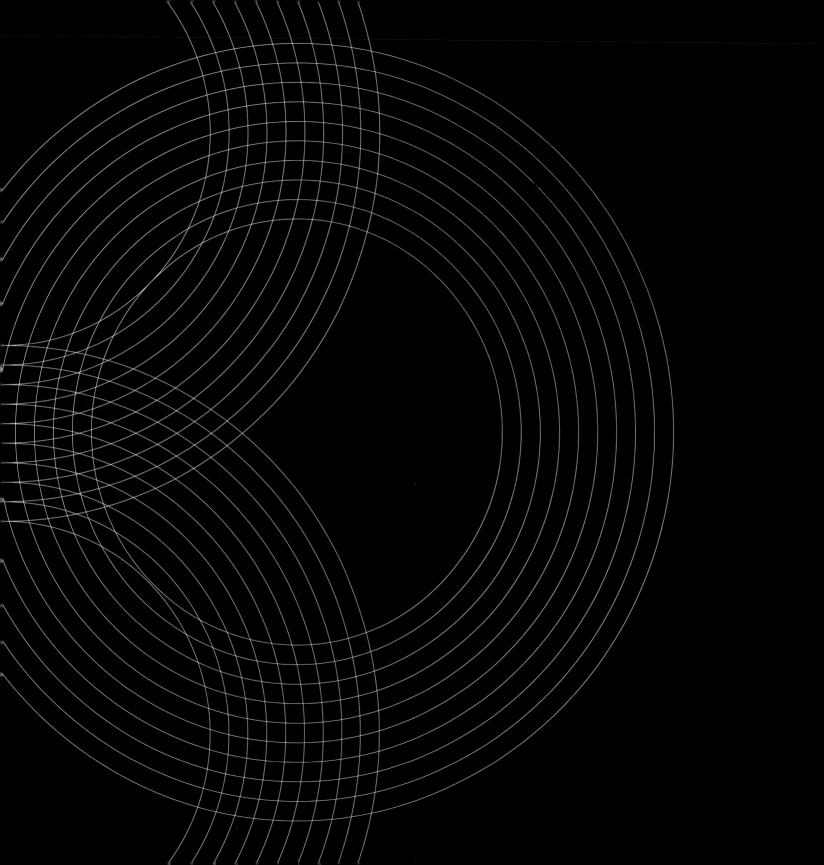

# Ginger Mule
# with Pistachio Vodka

Ginger and pistachios make for a surprising pairing. The resinous flavor in the pistachios works incredibly well with the pine flavor of ginger. The homemade pistachio vodka in this recipe is simple enough to concoct and leave to steep—and it is well worth the wait.

## GINGER SYRUP

4-inch piece fresh ginger, peeled and thinly sliced (about ½ cup)
2 cups water
1 cup sugar

Combine all the ingredients in a medium saucepot over medium-high heat and bring to a boil. Reduce the heat to a simmer and cook for 5 minutes, stirring occasionally, until the sugar has completely dissolved. Remove from the heat, cover, and let cool to room temperature. Strain into a clean glass bottle or jar, seal tightly, and store in the refrigerator for up to 4 weeks.

**Makes about 2 cups**

6 fresh mint leaves
¼ cup Ginger Syrup (recipe follows)
¼ cup fresh lime juice
¾ cup Pistachio Vodka (recipe follows)
About ½ cup ginger beer

In the bottom of a cocktail shaker, muddle the mint leaves with ginger syrup and lime juice. Add the vodka and fill with ice cubes. Shake well to chill. Strain into rocks glasses filled with 3 or 4 large ice cubes. Top each glass with ginger beer and serve.

## PISTACHIO VODKA

2 cups chopped toasted pistachios
3 cups vodka

Combine the chopped pistachios and vodka in a glass jar or bottle. Seal tightly and shake vigorously for 30 seconds. Store in a cool, dark place for 1 week, shaking vigorously once each day.

After a week, strain into a clean glass jar. Store in a cool, dark place for up to 3 months.

**Makes about 3 cups**

**SERVES 4**

**Main Subtypes:**

Wheat, barley, oats, farro, quinoa

**Best Pairings:**

Honey, citrus, roasted meat, butter, cheese, seafood, toasted nuts

**Surprising Pairings:**

Coconut, passion fruit, clam

**Substitutes:**

All varieties of grains may be substituted for one another. Also: lentils, split peas, dried beans

Grains are the edible seeds of grasses, and have been cultivated by humans as a food source for nearly 9,000 years. All grains are made up of three parts: bran, the edible outer seed covering; endosperm, the stored starch that makes up the bulk of the grain; and germ, which (like the yolk of an egg) is the embryo of the seed and has a high oil and nutrient content. Grain's flavor-creating aromatic compounds are mostly found in the bran and germ, which is why refined grain products like white flour are nearly flavorless until they undergo Maillard reactions during the cooking process. Compounds in the bran, when cooked, contribute roasted and toasted aromas (nuts and cocoa) while the germ has aromas of fat and can even smell slightly fishy. (While also technically a grain, rice has unique aromatic qualities that distinguish it from the other ingredients in this group. You can find rice, along with its own matrix, on page 210.)

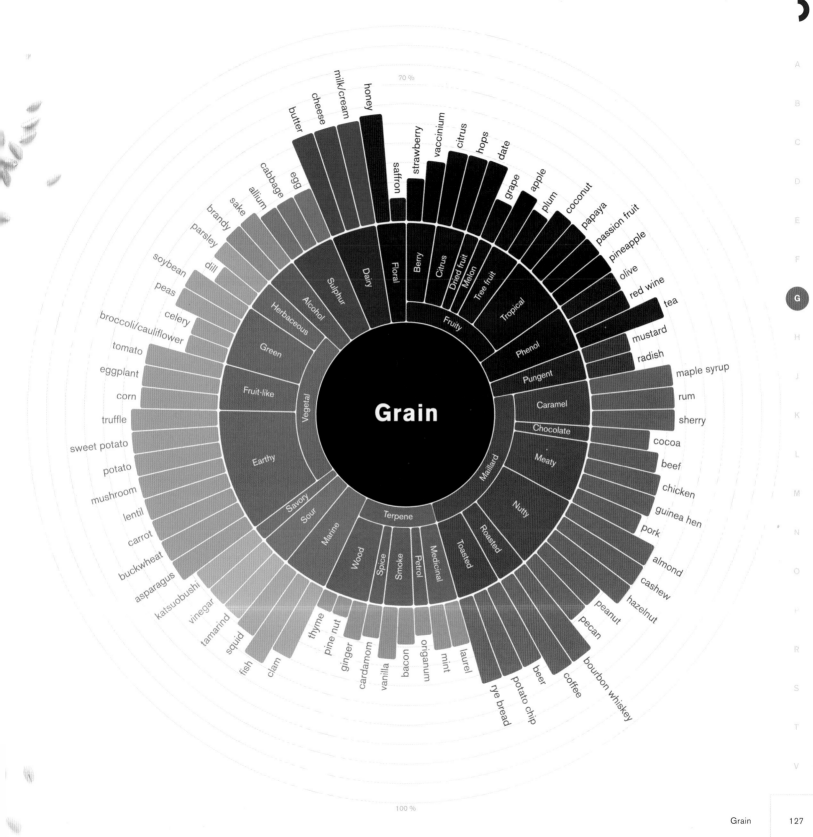

Grain

# Creamy Coconut Oats
# with Shrimp and Jalapeños

Coconut and oats may sound like something well-suited for the breakfast table or maybe the dessert course, but we thought this pairing needed to explore its savory side. Here, creamy oats get a spicy kick from jalapeño chile and are finished off with shrimp and toasted sesame seeds in a twenty-first-century twist on classic Southern shrimp and grits. All of the flavors in this dish build on the aromas naturally present in grains.

2 tablespoons unsalted butter

4 scallions (green and white parts), thinly sliced

2 cloves garlic, minced

1 teaspoon minced jalapeño chile

1 cup rolled oats

2 cups unsweetened coconut milk or one 13.5-ounce can
    coconut milk plus ⅓ cup water

16 medium shrimp, peeled, deveined, and grilled
    or sautéed

Fresh cilantro sprigs, for garnish

Toasted sesame seeds, for garnish

Melt the butter in a saucepot over medium heat. Add the scallions, garlic, and jalapeño and sauté until tender and very aromatic, about 3 minutes. Add the oats and stir well to coat them with the aromatics. Stir in the coconut milk and bring to simmer. Cook at a simmer until tender and creamy, stirring occasionally, about 15 minutes. If the mixture becomes too thick, add water and continue cooking to desired consistency.

Divide among four plates or shallow bowls and top with the shrimp. Garnish with cilantro and sesame seeds and serve immediately.

**SERVES 4**

A B C D E F G H J K L M N O P R S T V

**Best Pairings:**

   Arugula/cress, walnuts, mustard, vinegar, honey, stone fruit, cilantro, Sichuan peppercorns

**Surprising Pairings:**

   Pumpkin, capsicum, beet

**Substitutes:**

   Melon, figs, gooseberries, strawberries

While many varieties of grapes are grown for wine making or for drying into raisins, most of the grapes that turn up in stores and markets are table grapes, grown to be consumed fresh or cooked. Different varieties of table grapes may be grown to specific sizes, shapes, or colors, but all share certain flavor characteristics. The compound hexanol is a major contributor to grape flavor; it is also a key aroma in pomegranates and green vegetables such as green beans and lettuce. Other prominent aromas in grape flavor are berry and citrus. Grapes have an affinity for sour flavors, spices with citrus notes (like cilantro and Sichuan peppercorns), and fruit-like vegetables (bell peppers, pumpkin, and tomato).

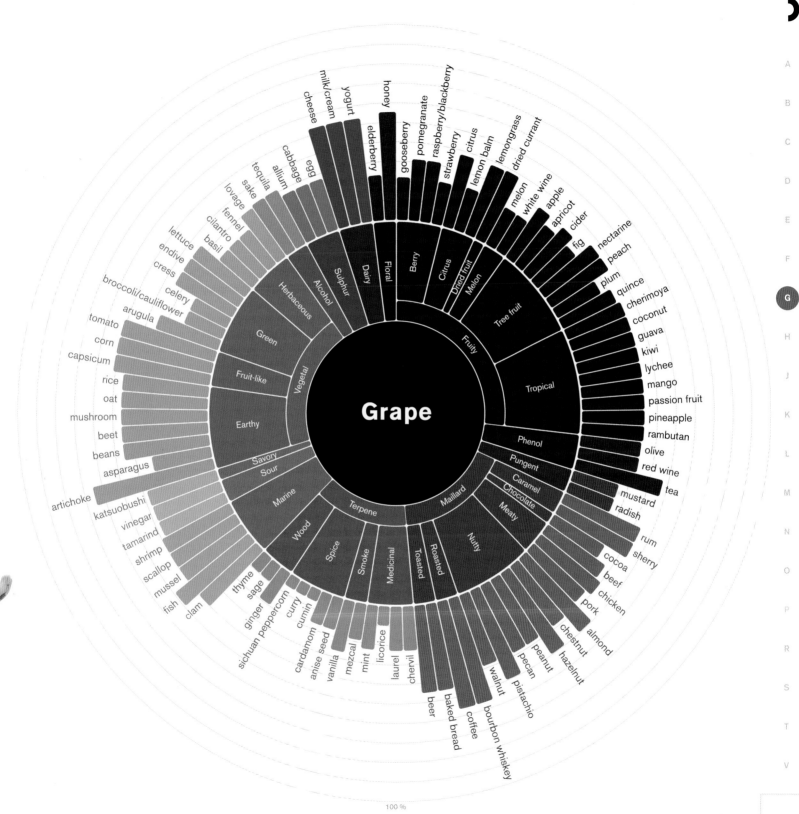

Grape

# Spice-Roasted Grapes

The herbs and spices here accentuate some of the lesser flavors in grapes (citrus and wood), while vinegar takes down the fruit's sweetness to give it a savory edge. Serve with sautéed fish, on a cheese plate, in a salad, or with roasted vegetables such as pumpkin, beets, or cauliflower, or roasted meat.

2 tablespoons extra-virgin olive oil
1 shallot, thinly sliced
1 teaspoon fennel seeds
1 teaspoon coriander seeds
3 pods cardamom
1 dried red chile (arbol or Thai) or
   ½ teaspoon red pepper flakes
6 to 8 branches fresh thyme
3 cups red or green grapes, cut in half
Kosher salt and freshly ground black pepper or ground
   Sichuan pepper
¼ cup red or white wine vinegar

Preheat the oven to 425°F.

Heat the olive oil in an ovenproof sauté pan over medium heat. When hot, add the shallots and sauté for 1 minute, until just softened. Add the fennel and coriander seeds, cardamom, chile, and thyme and cook 30 seconds more, until fragrant. Add the grapes, season well with salt and pepper, and toss well. Transfer the pan to the oven and roast for 5 minutes.

Remove the pan from the oven and place over medium heat on the stovetop. Add the vinegar and cook just until it is lightly reduced and syrupy. Toss well and serve warm.

**MAKES ABOUT 3 CUPS**

There are over 130 varieties of green beans, which—contrary to their name—range in color from green to yellow, purple, red, or streaked. (During cooking, red and purple varieties tend to revert to a dark shade of green.) Green beans are classified by how they grow: Bush beans grow on small plants and stay low to the ground, while pole beans grow tall and "climb," requiring a trellis or other support. A single variety of green bean may have both bush and pole variants. Green beans exhibit classic vegetal/green aromas produced by the alcohol hexanol, as well as much lighter, fruity aromas produced by a similar aldehyde called hexanal.

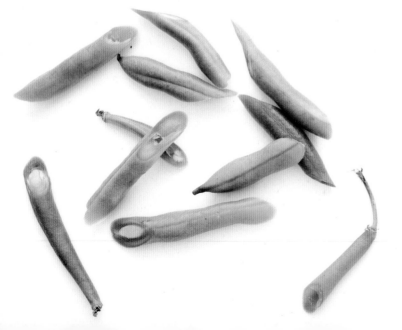

**Best Pairings:**

Fennel, mushroom, toasted nuts, alliums, wine, cheese, butter

**Surprising Pairings:**

Vanilla, pineapple, tea

**Substitutes:**

Green peas, edamame, fava beans, asparagus, cactus

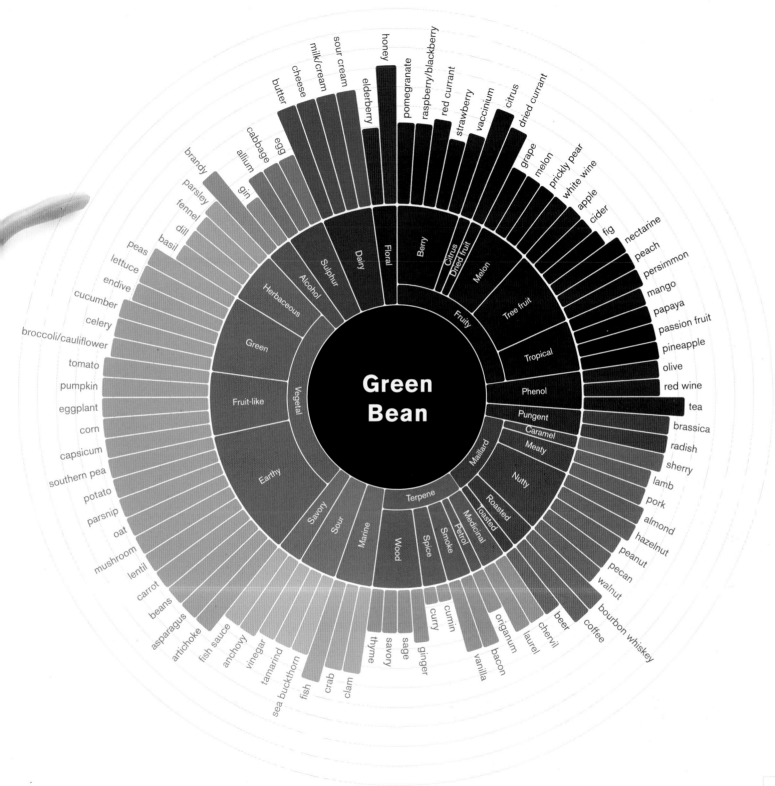

Green Bean

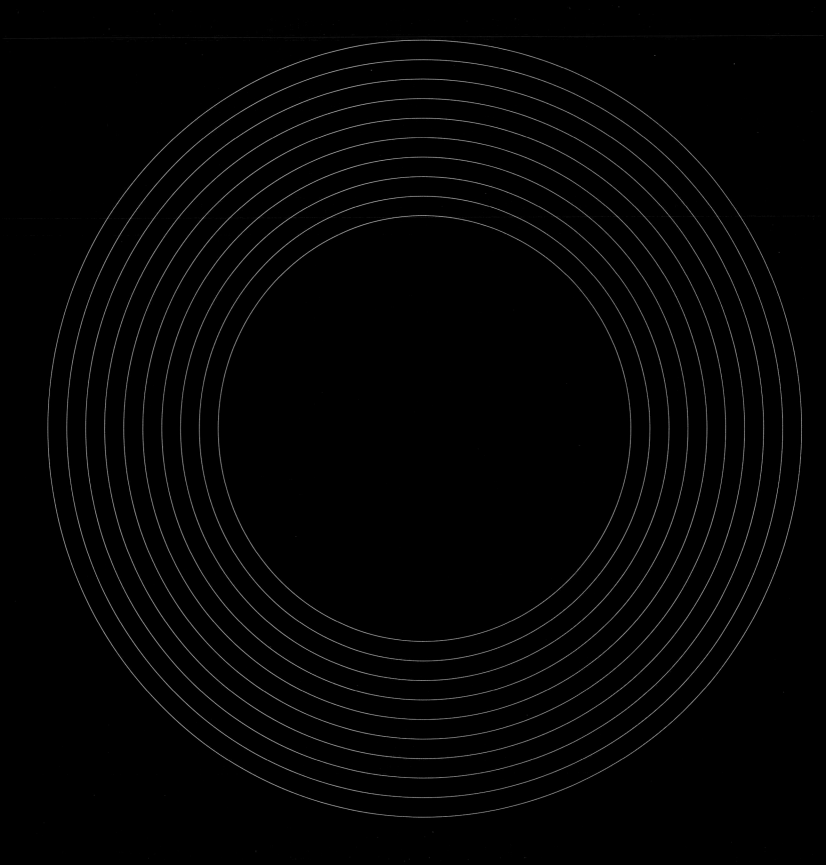

# Crunchy Green Bean "Granola"

Vanilla is a particularly tasty pairing for green beans. There are historical recipes from Venice that combine green beans with tomato and vanilla. In this recipe, green beans lean to the sweet side and lend crunch to a "granola" that makes a fantastic topping for dishes savory or sweet. To enhance this pairing even more, try the mixture as a garnish on puréed pumpkin or squash soup.

2 quarts water

2 tablespoons kosher salt

1 tablespoon sugar

1 pound thin green beans (haricots verts), trimmed

Vegetable oil or nonstick cooking spray

2 tablespoons unsalted butter

1 cup blanched hazelnuts

¼ cup unsweetened shredded coconut

2 tablespoons white sesame seeds

¼ cup honey

1 teaspoon pure vanilla extract

Preheat the oven to the lowest possible setting, between 150 and 200°F, or a food dehydrator to 130°F.

Bring the water, salt and sugar to a boil in a medium saucepot. Add the green beans and cook for 1 minute, until barely tender. Drain the beans in a colander; shake occasionally to drain more. Spread the beans in a single layer on a baking sheet or dehydrator tray and let sit until dry and cooled.

Brush or spray the beans with a very light coating of oil. Place in the oven or dehydrator to dry. At 200°F, the beans will take about 3 hours; at150°F, the beans will take about 6 hours; at 130°F, they will take about 10 hours.

When the beans are completely dehydrated, coarsely chop them and set aside in an airtight container.

Preheat the oven to 250°F. Line a rimmed baking sheet with parchment paper and brush or spray it lightly with oil.

Melt the butter in a small saucepot over medium heat. Add the hazelnuts, coconut, and sesame seeds and stir well to coat. When all of the nuts are well toasted and fragrant, stir in the honey. Continue cooking until the honey begins to simmer, then cook 1 minute more. Remove from the heat and stir in the vanilla. Immediately transfer the mixture to the prepared baking sheet and press out to thin layer with a silicone spatula. Bake for 30 minutes, until well toasted.

Remove from the oven and let cool completely. Coarsely chop the nuts, mix in the green beans, and serve. Or store in an airtight container at room temperature for up to 3 weeks.

**MAKES 2 CUPS**

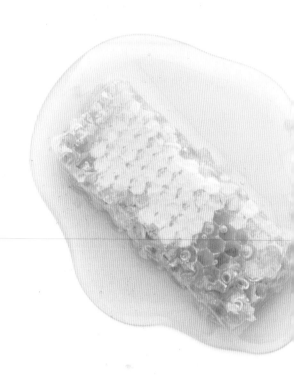

From the spiced flavor of clover honey to the citrusy flavor of orange blossom honey, honey can vary quite widely depending on where the bees that created it sourced their nectar. Many commercial honeys are blended, with a mixture of honeys from multiple floral sources. All varieties of honey share certain flavor characteristics, however. For example, every honey contains Maillard-like aromas created by furanones, which also have strong antimicrobial properties. Floral aromas are persistent throughout all varieties of honey, as well, as are vegetal aromas. Honey is an extremely versatile sweetener, and generally pairs well with all ingredients, although like any food, it has some ingredients that it *really* gets along with.

**Best Pairings:**

Citrus, dried fruit, wine, mustard, tea, lettuce/greens, vinegar, alcohol

**Surprising Pairings:**

Olive, sage, capsicum

**Substitutes:**

Maple syrup, molasses, cane syrup, light treacle

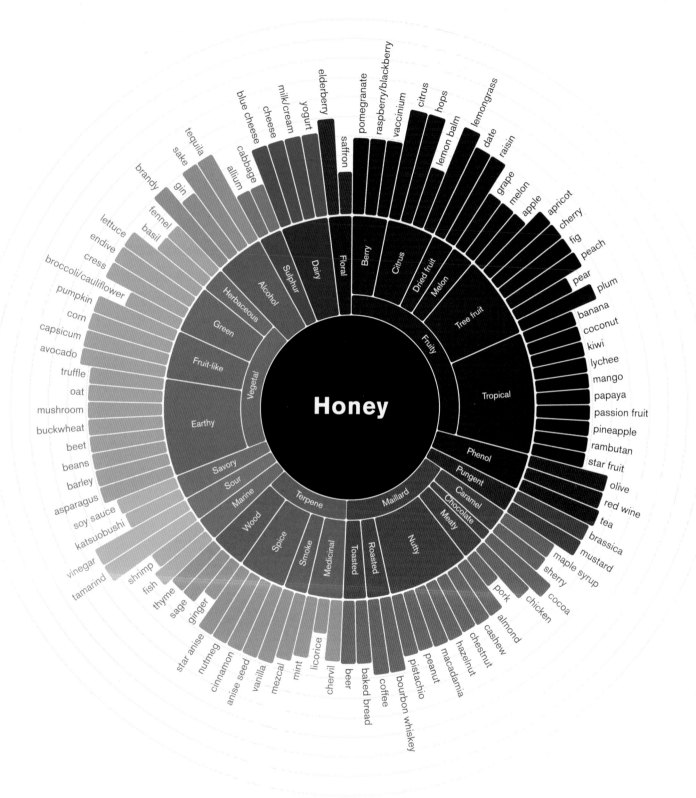

Honey

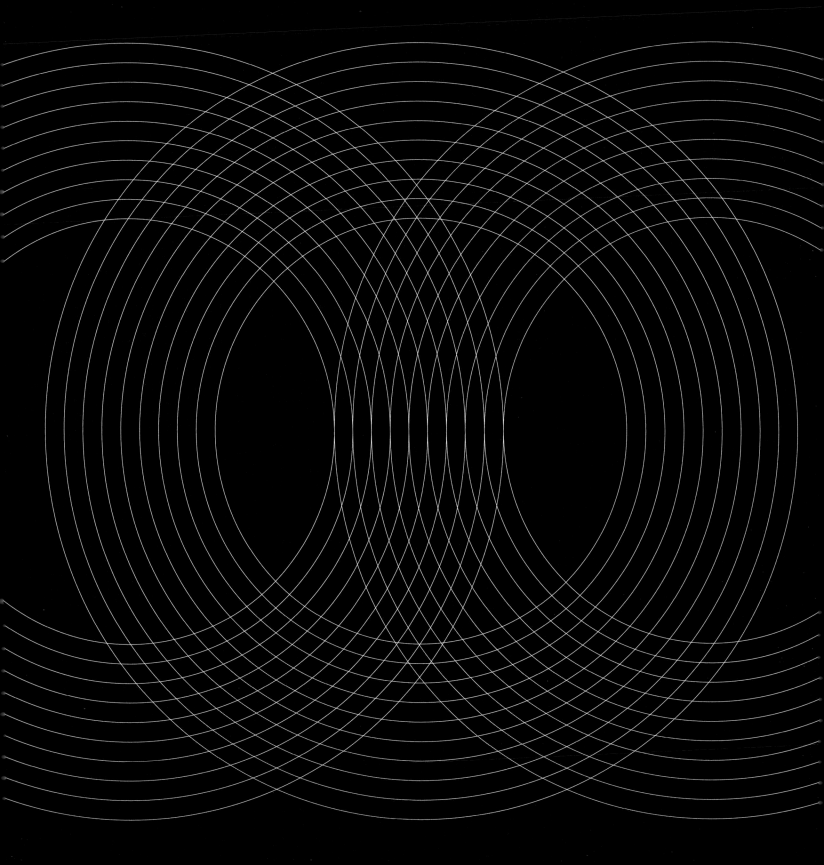

# Spicy Honey Syrup

With its affinity for alcohol, honey is a particularly good option for sweetening and flavoring cocktails. This spicy honey syrup is particularly great in tequila, gin, and vodka drinks. For a nonalcoholic beverage, use this syrup to sweeten lemonade.

1 cup honey
1 cup water
1 jalapeño, serrano, or other hot chile, thinly sliced
Grated zest of 1 lime
4 branches fresh thyme
1 teaspoon black or Sichuan peppercorns
1 pod star anise

Combine all the ingredients in a small saucepot and bring to a simmer. Cook at a low boil for 5 minutes. Remove from the heat and allow to cool completely in the pot. Strain into a clean glass bottle or jar, seal tightly, and store in the refrigerator for up to 1 month.

**MAKES 2 CUPS**

**Best Pairings:**

Lemon, orange, mushrooms, roasted meat,
alliums, cream, butter

**Surprising Pairings:**

Cherry, gin, crab, vanilla, fennel, dill, cilantro

**Substitutes:**

Crosnes, jicama, water chestnuts, potato

Jerusalem artichokes, also known as sunchokes, are
the starchy root of a plant in the sunflower family.
While their flavor is often described as being similar to
that of cooked artichokes (hence their name), modern
Jerusalem artichokes have a richer, nuttier flavor,
with woody, pine, and toasted nut scents. Jerusalem
artichokes are extremely versatile, and may serve as a
more flavorful substitute for potatoes in any recipe that
calls for them. Jerusalem artichokes do not need to be
peeled before being eaten, although a good scrub first
is advisable, to remove any grit from the skin. When
thinly sliced, raw Jerusalem artichokes can add a
delicious, nutty crunch to salads.

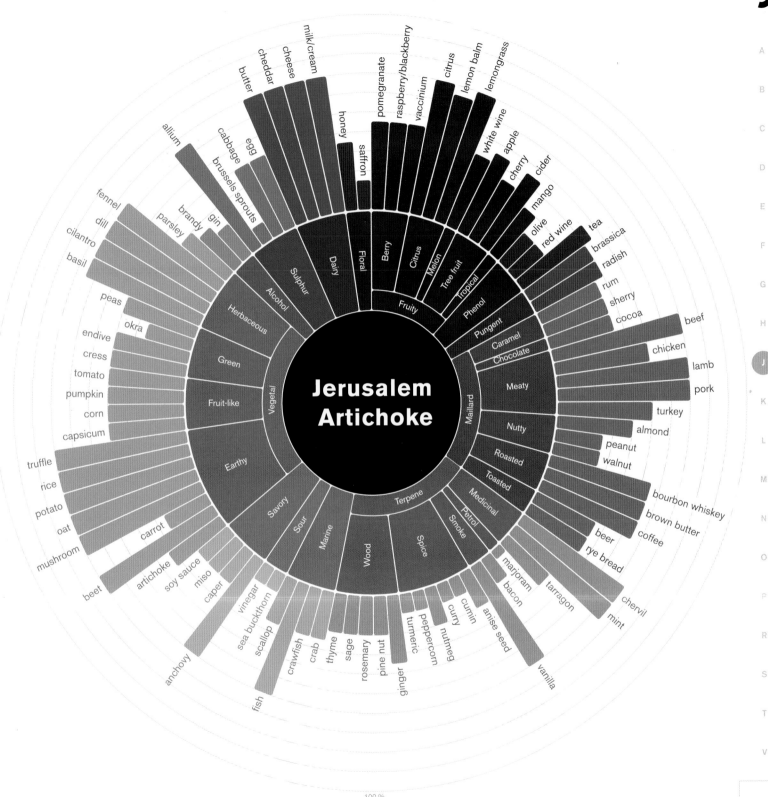

Jerusalem Artichoke

# Gin and Brown Butter Emulsion

The juniper and botanical flavors of gin come together with the Maillard richness of brown butter to make the perfect match for drizzling over roasted Jerusalem artichokes. But this combo is great with other main dishes, too. Make it a regular part of your repertoire; it's great on roasted meat, potatoes, fish, or other root vegetables.

½ pound (2 sticks) unsalted butter

4 cloves garlic, peeled and smashed

4 to 6 branches woody herbs (such as sage, thyme, rosemary, or oregano)

¼ cup minced shallot

½ cup gin

2 tablespoons heavy cream

Kosher salt and freshly ground black pepper

Fresh lemon juice

Melt the butter in a small saucepot over medium heat. Bring to a simmer and cook until the milk solids have settled to the bottom of the pot, all the water has evaporated (the bubbling will have stopped), and the butter looks clear. Continue cooking until the butter takes on a deep brown color and has a nutty, roasted aroma. This can take as long as 15 minutes. Remove from the heat and carefully add the garlic and herbs to the butter. Set aside to infuse and cool.

Transfer a tablespoonful of the browned butter to a small sauté pan. Add the shallots and sauté over medium heat until tender and sweet smelling. Remove the pan from the heat and add the gin. Carefully place the pan back on the heat, bring to a boil, and reduce by half. Stir in the heavy cream and reduce 1 minute more, until the mixture is reduced by half. Remove from the heat.

Strain the butter through a fine-mesh sieve lined with damp cheesecloth into a container with a spout, such as a heatproof glass measuring cup.

Transfer the reduced gin mixture to a blender and turn on the machine at low speed. Pour the brown butter into the blender in a slow, steady steam while the machine is running. When all the butter has been added, the sauce should be emulsified and thickened. Season to taste with salt, pepper, and lemon juice. Use immediately.

**MAKES ABOUT 1¼ CUPS, ENOUGH FOR 8 SERVINGS**

While the kiwi fruit is associated with New Zealand, the fruit is actually native to China—which is why, in Europe, it was once called the Chinese gooseberry. True to that name, the kiwi is technically a berry. The tender green or golden flesh has a flavor that is unique among all fruits, coming from a mix of vegetal/green and fruity (apple, pineapple, and berry) aromas. Kiwis also contain actinidin, an enzyme that works as a natural tenderizer for meat. Rub a cut kiwi directly onto raw meat and then apply the marinade of your choice; the actinidin in the fruit's flesh will immediately go to work breaking down connective tissue. (This trick works best for short marinade times, as actinidine can make meat mushy after 24 hours.)

**Best Pairings:**

Orange, lime, vaccinium, strawberries, wine, brandy

**Surprising Pairings:**

Basil, cilantro, jalapeño, vaccinium

**Substitutes:**

Honeydew, fig, grapes

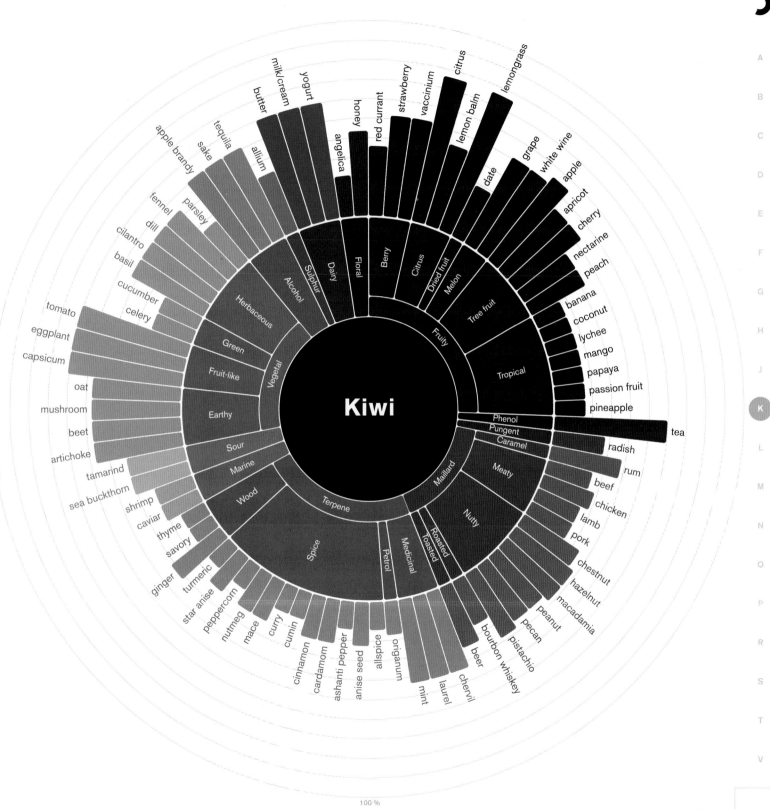

Kiwi

# SunGold Kiwi and Cranberry Relish

The growing seasons of cranberries and kiwis just barely overlap—but thank goodness they do. The flavors of these two ingredients blend together perfectly, creating a truly unique combination of fruit, spice, and tannins. Cilantro and jalapeño add extra intrigue to the dish, rounding out its flavor profile perfectly. While most of the kiwis you see are of the green, fuzzy-skinned Hayward variety, this recipe calls for the yellow, smooth-skinned SunGold cultivar, whose color complements the bright red of cranberries. Serve with any grilled or roasted meat.

1 cup sugar
2 cups water
Zest of 1 orange, removed in strips with
    a vegetable peeler
One 2-inch piece ginger, peeled and thinly sliced
3 cups fresh or frozen cranberries
1 cup diced peeled SunGold or green kiwis
    (about 4 kiwis)
½ jalapeño chile, minced
1 tablespoon minced fresh cilantro
Kosher salt and freshly ground black pepper

Combine the sugar, water, orange zest, and ginger in a small saucepot and bring to a boil. Reduce the heat to simmer and cook 2 minutes, until the sugar has dissolved completely. Add the cranberries and cook until they all burst, but have not completely broken down, 4 to 5 minutes.

Drain the cranberries and reserve the liquid for another use (like cocktails!). Pick out and discard the ginger and orange zest.

Transfer the cranberries to a bowl and let cool to room temperature. Fold in the diced kiwi, jalapeño, and cilantro. Season to taste with salt and pepper. Serve immediately, or cover the bowl and store in the refrigerator of up to 3 days.

**MAKES 2 CUPS**

**Best Pairings:**

Toasted breadcrumbs, lemon, orange, mustard, soft herbs, butter, tea

**Surprising Pairings:**

Shrimp, strawberry, mango, sauerkraut

**Substitutes:**

Beef, game meat

Lamb—and its older relative, mutton—is the meat derived from domesticated sheep. To be classified as lamb, a sheep may be up to one year old at the time of harvest; to count as mutton, it is typically over two years of age. The meat of lamb is somewhat more tender and lighter in color than mutton, and has a more delicate flavor. Both are red meats that look very similar to beef, although with a distinct flavor thanks to branched-chain fatty acids that develop naturally in sheep. These fatty acids have strong aromas—often described as sweaty, barnyardy, fatty, or sour—that some people find unappealing. Because these compounds increase with the age of the animal, most consumers prefer the taste of lamb to mutton. If you're a sheep-shy type, know this: Because the majority of lamb's flavor is found in fat, lean cuts from the loin have the mildest flavor.

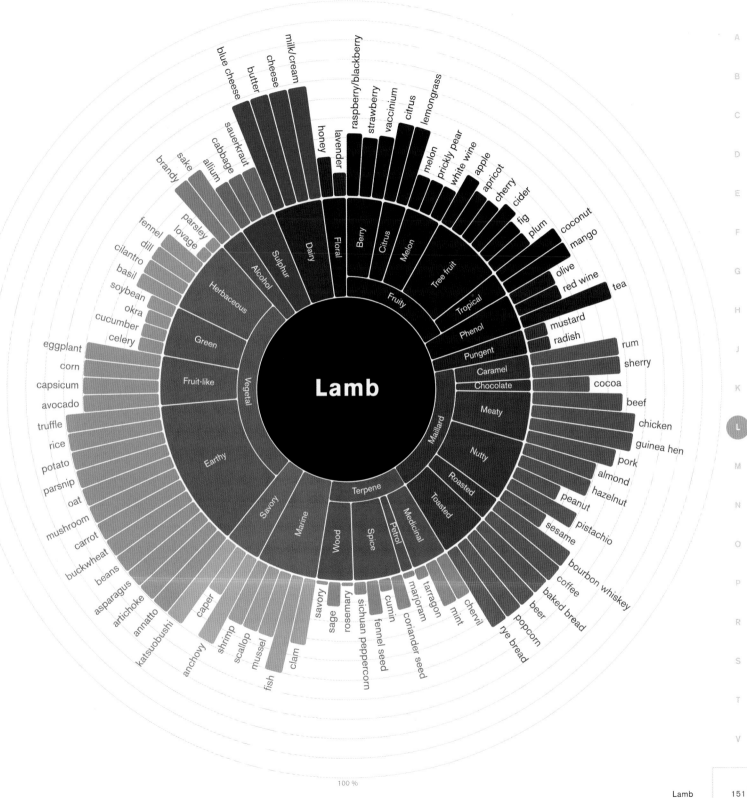

# Lamb Seasoning Rub

Lamb pairs best with Maillard and earthy flavors, which coffee and mushroom powder help amplify in this seasoning. Fennel and mint, whose flavors are more classically paired with lamb, round out the mix. Rub this on meat before grilling, roasting, or cooking sous vide.

2 tablespoons ground coffee
1 teaspoon fennel seeds
1 teaspoon dried onion flakes
Finely grated zest of 1 lemon
1 tablespoon dried mint, finely crumbled
1 teaspoon mushroom powder
¼ cup kosher salt
2 teaspoons freshly ground black pepper

Combine the coffee, fennel seeds, onion, and lemon zest in a mortar and grind with a pestle until well mixed but not a fine powder, or grind them together in a spice grinder. Add the mint, mushroom powder, salt, and pepper and mix well. Transfer to a glass jar or plastic container, seal tightly, and store in a cool, dark place for up to 1 month.

**MAKES ABOUT ½ CUP**

**Best Pairings:**

Cardamom, cinnamon, peppercorn, rosemary, citrus, soft herbs, alcohol

**Surprising Pairings:**

Cauliflower, rosemary, caraway

**Substitutes:**

Citrus zest, cilantro, ginger, lemon balm, lemon verbena

Lemongrass provides a distinctive flavor to the cuisines of Southeast Asia and India. Fresh lemongrass has bright lemon and fruity aromas, and it is often found in recipes alongside chiles, coconut, and lime—all great pairings for the plant's dominant citrus aromas. Other prominent aromas in lemongrass include wood and spice. It's important to note that lemongrass only contributes aromas to a dish, as its stalk is woody and tasteless. To use lemongrass, slice or mince it extremely fine to make it more tender. Or leave it in large pieces, "bruise" it with the back of knife, and add it to a dish during cooking, then remove it before serving.

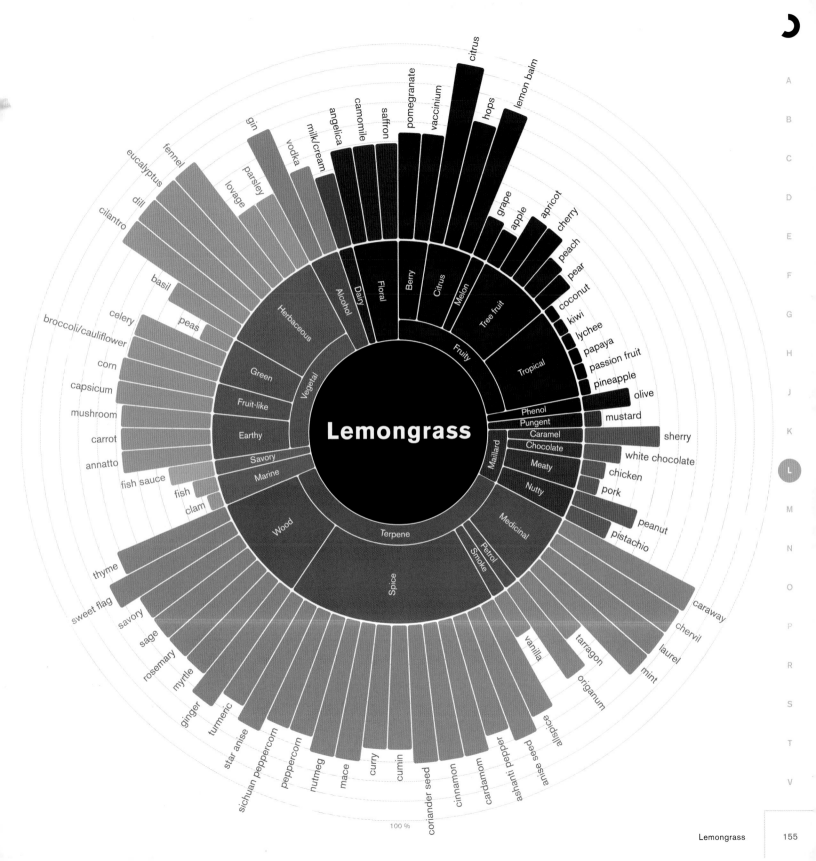

Lemongrass

# Lemongrass Bitters

These lemongrass bitters are delicious in cocktails, but you can also use them to up your baking game: Add the bitters to any recipe that calls for vanilla extract, in addition to or in place of the vanilla. They are especially good in chocolate desserts and cookies.

1 stalk lemongrass
8 pods cardamom
1 cinnamon stick, crushed
1 tablespoon coriander seeds
4 branches fresh rosemary
1 cup grain alcohol or high-proof vodka

Trim the root end from lemongrass and discard. Cut 3 inches of the pale stalk starting from the trimmed end, and slice crosswise as thin as possible. (Reserve the remaining lemongrass stalk and leaves for another use.) Combine the lemongrass, cardamom, cinnamon, coriander, and rosemary in a glass jar. Add the alcohol and seal tightly. Shake vigorously for 30 seconds. Store in a cool, dark place for 3 weeks, shaking the jar vigorously every couple of days to promote infusion.

After 3 weeks, strain the bitters through cheesecloth into a clean glass jar or bottle. Seal tightly and store in a cool, dark place. Bitters will keep almost indefinitely.

**MAKES ABOUT 1 CUP**

Lettuces were first cultivated as a food source for humans by the ancient Egyptians. While lettuce can add color, texture, and bulk to a meal, it has very little flavor. Most lettuces contain a mere twenty aromatic compounds (strawberries have 400-plus). Lettuce flavor can be increased by cooking, such as grilling romaine or sautéing leaf or butter lettuce. Heat will also cause lettuce to wilt, however, meaning that you pay for that increased flavor with a decrease in texture.

**Best Pairings:**

Vaccinium, apple, mushrooms, mint, buttermilk, yogurt, fish

**Surprising Pairings:**

Tamarind, coffee, tea, rhubarb

**Substitutes:**

Cucumber, other greens, cress, brassicas

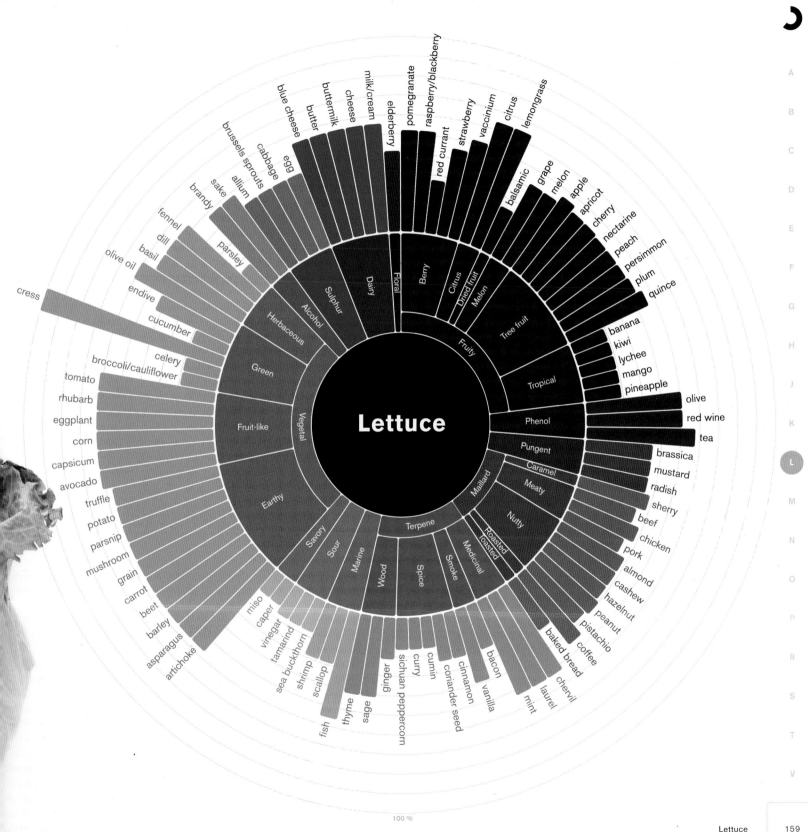

**Lettuce**

Floral
Dairy
milk/cream
cheese
buttermilk
butter
egg
blue cheese
cabbage
brussels sprouts
allium
sake
brandy
fennel
dill
basil
olive oil
endive
cucumber
celery
broccoli/cauliflower
tomato
rhubarb
eggplant
corn
capsicum
avocado
truffle
potato
parsnip
mushroom
grain
carrot
beet
barley
asparagus
artichoke
miso
caper
vinegar
tamarind
sea buckthorn
shrimp
scallop
fish
thyme
sage
ginger
sichuan peppercorn
cumin
curry
coriander seed
cinnamon
vanilla
bacon
laurel
mint
chervil
baked bread
coffee
pistachio
peanut
hazelnut
cashew
almond
pork
chicken
beef
sherry
radish
mustard
brassica
tea
red wine
olive
pineapple
mango
lychee
kiwi
banana
quince
plum
persimmon
peach
nectarine
cherry
apricot
apple
melon
grape
balsamic
lemongrass
citrus
vaccinium
strawberry
red currant
raspberry/blackberry
pomegranate
elderberry
cress

Sulphur
Alcohol
Herbaceous
Green
Fruit-like
Vegetal
Earthy
Savory
Sour
Marine
Wood
Spice
Medicinal
Smoke
Toasted
Roasted
Nutty
Meaty
Caramel
Maillard
Pungent
Phenol
Tropical
Tree fruit
Fruity
Melon
Dried fruit
Citrus
Berry

Terpene

A
B
C
D
E
F
G
H
J
K
L
M
N
O
P
R
S
T
V

100 %

# Tamarind-Glazed Wedge Salad
## with Creamy Herb Dressing

This recipe amps up lettuce's flavor by lightly cooking it—but what really kicks the taste into overdrive is the spicy, sour, and sweet tamarind glaze applied to the lettuce before broiling, as well as the cooling herb dressing that finishes the dish.

1 head iceberg lettuce, cut into quarters, or 2 romaine
     hearts, cut in half lengthwise
Kosher salt
¼ cup tamarind concentrate
¼ cup rice vinegar
2 tablespoons light brown sugar
½ jalapeño chile, minced
2 cloves garlic, chopped
1 tablespoon minced ginger
Creamy Herb Dressing (recipe follows)

## GARNISHES

Crumbled cooked bacon
Diced tomatoes
Crumbled blue cheese or feta

Preheat the broiler to high.

Arrange the lettuce on a baking sheet, cut side up. Lightly season each piece with salt and set aside.

To prepare the glaze, combine the tamarind, vinegar, sugar, jalapeño, garlic, and ginger in a blender and purée until smooth.

Pat the lettuce wedges dry with a paper towel. Brush each with a generous coating of the glaze. Place the baking sheet on the center rack of the oven below the broiler. Broil for 1 to 2 minutes.

Remove the lettuce and apply a second coat of glaze. Place under the broiler for 1 minute more, until the glaze thickens and browns.

Divide the lettuce among four plates and top with dressing and garnishes. Serve immediately.

## CREAMY HERB DRESSING

1 cup buttermilk, or 1 cup whole milk plus 1 teaspoon
     white
     distilled vinegar
½ cup sour cream
1 tablespoon minced fresh chives
1 tablespoon minced fresh basil
1 tablespoon minced fresh dill
1 clove garlic, minced
1 teaspoon kosher salt
Pinch of ground cumin
Juice of 1 lemon
Worcestershire sauce
Tabasco sauce

Combine the buttermilk, sour cream, chives, basil, dill, garlic, salt, cumin, and lemon juice in a small bowl and whisk well to combine. Season to taste with Worcestershire and Tabasco. Cover and store in the refrigerator for up to 10 days.

**Makes about 2 cups**

**SERVES 4**

Melons belong to the Cucurbitaceae family, which also includes squash, pumpkin, zucchini, and cucumber. The most commonly eaten melons—watermelon, cantaloupe, and honeydew—grow in a variety of cultivars that can vary slightly in size, shape, and color. Aroma plays a major role in melon flavor, so much so that a ripe cantaloupe can be smelled through its rind. Give the whole melon a squeeze, too: A slight give to the flesh can also indicate ripeness in cantaloupe and honeydew. Ripe watermelons, on the other hand, have a hollow sound and a pronounced pale yellow spot on their bottom. Melons reach ripeness in late spring through summer. While melons' specific flavors vary, they are all defined by fruity aromas.

**Main Subtypes:**

Watermelon, cantaloupe, honeydew

**Best Pairings:**

Mushrooms, vanilla, toasted nuts, mustard, ham/bacon, wine

**Surprising Pairings:**

Clam, olive, corn, mustard

**Substitutes:** Apricot, grapes, berries

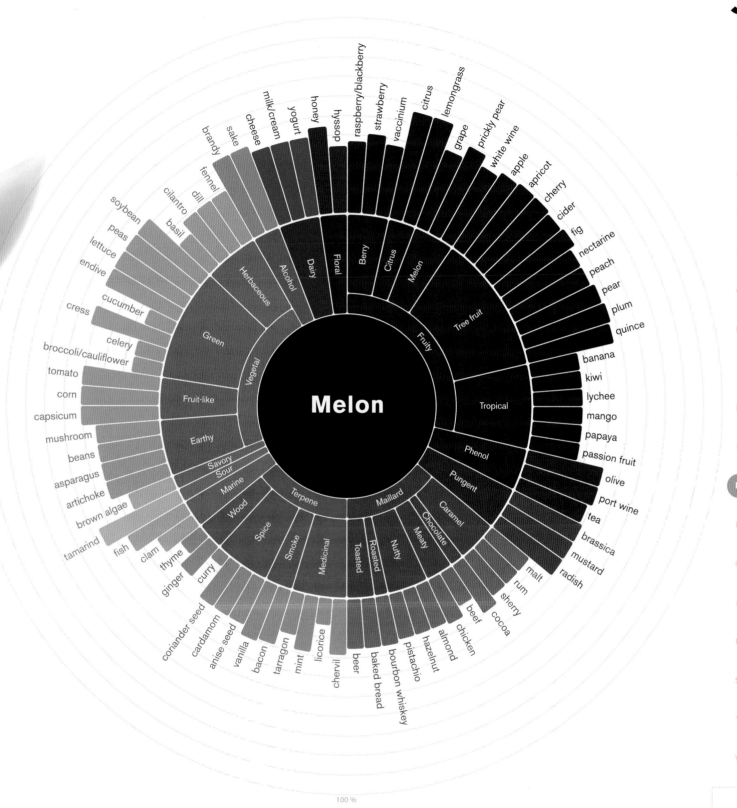

Melon

# Bacon Tapenade and Melon Bruschetta

Any dish that puts bacon on a bruschetta is almost guaranteed to be good. This meaty, smoky tapenade makes the ultimate pairing for honeydew melon and a thin, crisped slice of sourdough—but it may also become your go-to sandwich spread or sophisticated cheese plate staple. If gluten-free is your thing, forget the bread and instead dollop the tapenade atop pieces of cubed watermelon.

8 strips thick-cut bacon

2 cups pitted niçoise olives

1 anchovy fillet, rinsed

1 clove garlic, peeled

3 tablespoons drained capers, chopped

2 hard-boiled eggs, grated

1 tablespoon chopped fresh parsley

Extra-virgin olive oil

Fresh lemon juice

Freshly ground black pepper

12 slices sourdough bread

Honeydew or cantaloupe, peeled, seeded, and thinly sliced or shaved into strips with a vegetable peeler

Cook the bacon until crisp. Drain on paper towels. Set aside the fat for later, reserve for another use, or discard. Crumble the bacon.

Combine the olives, anchovy, and garlic in a food processor and pulse until no large pieces remain; take care not to process the mixture so much that it turns into a smooth paste. Transfer to a bowl and stir in the capers, eggs, and parsley. Stir in olive oil until the mixture is spreadable but still thick, about 2 tablespoons. Season to taste with lemon juice and freshly ground black pepper. Gently mix in the bacon crumbles.

Heat a grill for direct cooking or preheat the oven to 450°F. Arrange the bread in one layer on a baking sheet. Brush the slices on one side with olive oil or bacon fat. Toast directly on the grill until lightly charred, or in the oven until golden and crisp.

Spread some of the tapenade on each piece of bread and top with ribbons of melon. Serve immediately.

**SERVES 6**

**Main Subtypes:**

Snail, mussel, oyster, cockle, scallop, clam, octopus, squid

**Best Pairings:**

Mushrooms, grains, nuts, breadcrumbs, cream, butter, beet, asparagus

**Surprising Pairings:**

Vanilla, rutabaga, eggplant

**Substitutes:**

Any mollusk may be substituted for another; also: crustaceans such as shrimp and crab

There are 50,000 known species of mollusks, ranging from clams to octopuses, but while they can look dramatically different from one another they have many underlying similarities. Like most animals, each type of mollusk has a specific flavor that is affected by its environment, an influence that—as far as mollusks are concerned—is most pronounced in filter feeders such as oysters, mussels, scallops, and clams, which siphon water through their bodies to filter out the plankton and algae they feed on. Generally speaking, mollusks contain fewer of the nitrogen and sulfur compounds that create "fishy" aromas in lean and fatty fish; the majority of their flavor is created by earthy and Maillard aromas, making earthy, vegetal, and roasted flavors the best pairings. That said, when deciding which type of mollusks to use in a recipe, it is important to also consider taste and texture: Are the mollusks you're using briny or sweet, tender or chewy? A sweet scallop will pair nicely with smoky vanilla; a briny oyster, not so much.

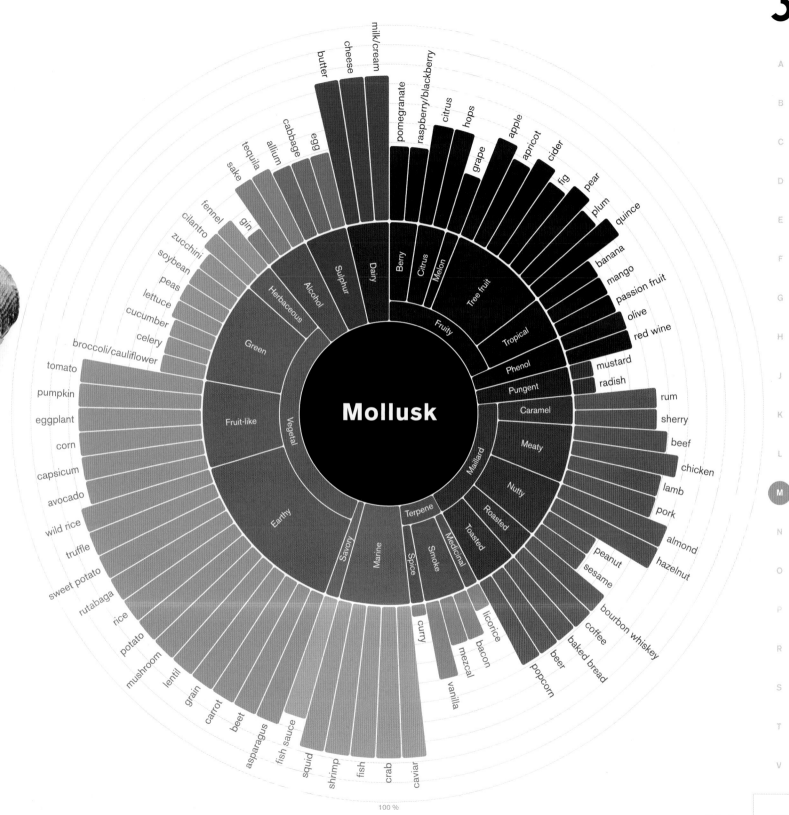

Mollusk

# Barley and Cauliflower "Risotto"

This "risotto" is redolent of the heady aromas of anchovy, cumin, and garlic, and makes a delicious base for sautéed, roasted, or grilled mollusks of all types, from clams to scallops to octopus. Alternatively, think of the crunchily dried variation as the gluten-free breadcrumbs of the future. One spoonful of this concoction is the ultimate topping for mollusks before baking or broiling.

2 cups pearled barley
8 cups water
1 tablespoon kosher salt
1 tablespoon unsalted butter
1 tablespoon extra-virgin olive oil
2 cloves garlic, minced
2 anchovy fillets, chopped
½ teaspoon ground caraway seed or cumin
1 head cauliflower (about 1½ pounds), stems and
    florets grated
½ cup dry white wine
Kosher salt and freshly ground black pepper
2 tablespoons minced fresh parsley
Grated zest of ½ lemon

Combine the barley, water, and salt in a medium saucepot. Bring to a boil over medium-high heat. Reduce the heat to a simmer, cover the pot, and cook until the barley is tender, 30 to 40 minutes. Set a colander over a bowl and drain the barley; set aside the cooking liquid.

In a separate saucepot, melt the butter with the olive oil over medium heat. Add the garlic, anchovies, and caraway and cook until the garlic is sizzling and fragrant.

Stir in the cauliflower and turn the heat to high. Cook, stirring occasionally, until contents of the pot begin to sizzle again. Add the wine and continue cooking until it has evaporated completely, about 2 minutes. The cauliflower should be tender at this point. If not, add a little of the barley cooking water and simmer until it is.

When the cauliflower is tender, stir in the barley and season to taste with salt and pepper. Add barley cooking water until the mixture has the consistency of a creamy risotto. Rewarm over low heat, if necessary, stirring frequently. Divide among four shallow bowls, garnish with the parsley and lemon zest, and serve immediately.

## VARIATION

Immediately after adding the barley and without adding extra liquid, spread the cooked mixture on a baking sheet to cool and dry. Stir in minced fresh herbs such as parsley or thyme and grated cheese, and use as a topping for shellfish. Spoon some into open bivalves (clams, oysters, or mussels), drizzle with olive oil, and broil until crisp. Store any remaining "crumbs" in an airtight container in the refrigerator for up to 10 days.

**SERVES 4**

Edible mushrooms fall into two categories: cultivated and wild. Cultivated mushrooms are easily grown and readily available year-round. Their flavor is consistent, though it tends to be blander and less dynamic than that of wild mushrooms. Wild mushrooms have to be hand harvested or foraged, and their scarcity—as well as the demand for their bold flavors—makes them more sought after and expensive. The flavor of both types of mushroom comes from the compound 3-octanol, which has pronounced aromas of moss and soil with notes of citrus and nut. Although they're not typically perceptible, mushrooms also contain a number of fruity esters which make for very interesting pairings. Mushrooms have very little taste—unless you count umami, of which they are one of the highest natural sources.

**Best Pairings:**

Allium, parsley, origanum, thyme, lavender, dried and fresh fruit

**Surprising Pairings:**

Strawberry, coconut, tamarind, cocoa

**Substitutes:**

Most mushrooms may be substituted for each other; also: eggplant, zucchini/summer squash, tofu, miso, fish sauce, soy sauce

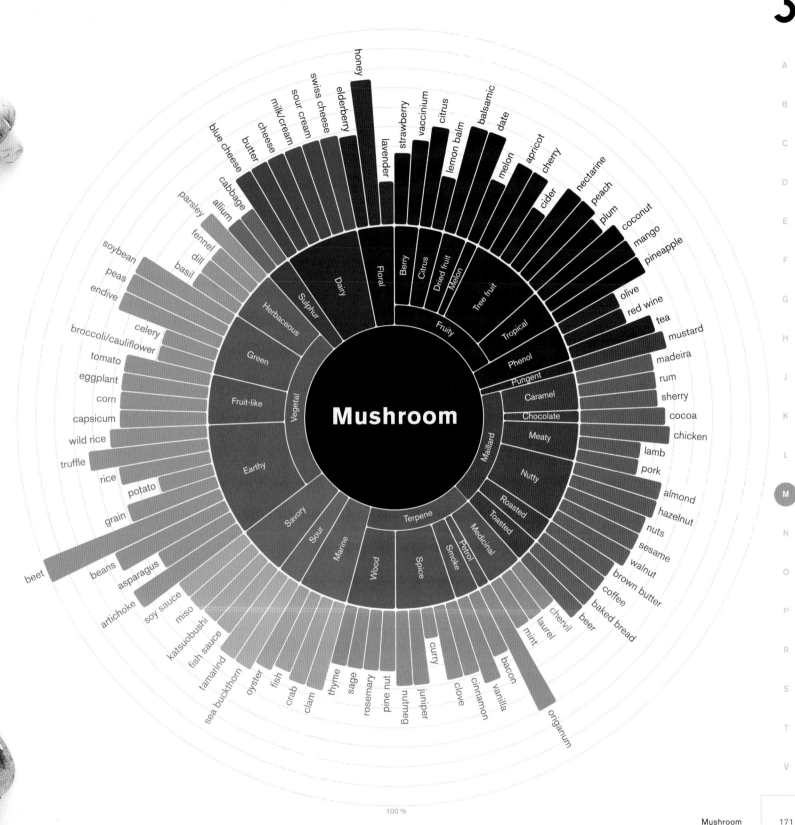

Mushroom

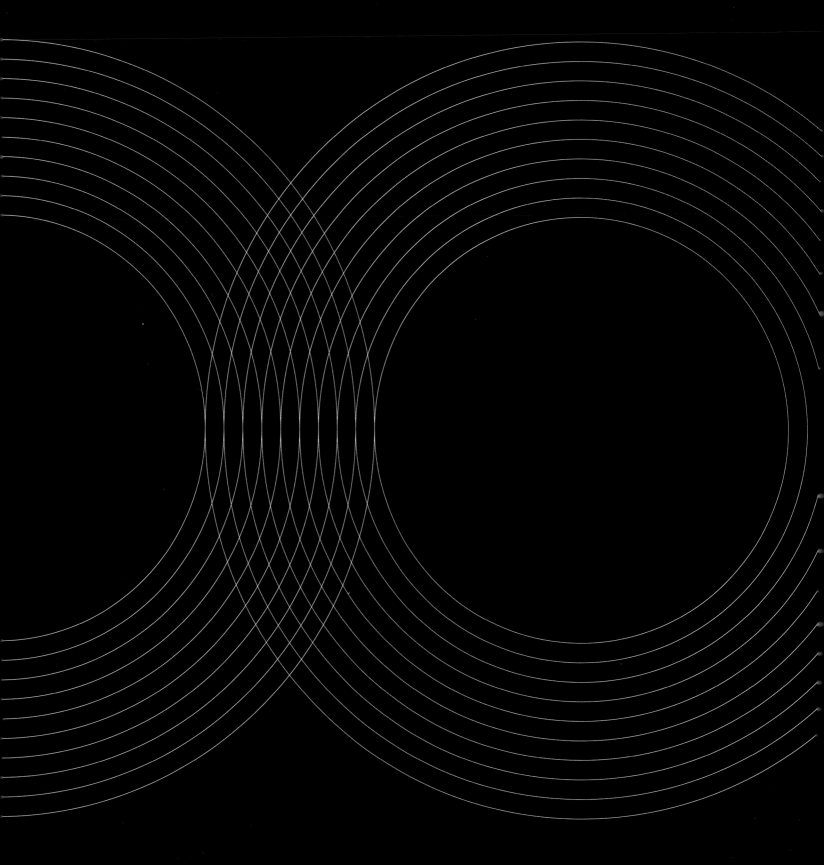

# Porcini, Hazelnut, and Chocolate Torte

This torte is basically a gluten-free brownie with an added burst of umami from a most unlikely source. Mushrooms and hazelnuts are very compatible, chemically speaking, and while the former is less typical in a dessert course than the latter, they turn out to make great partners in sweetness. Porcini powder both adds umami and helps accentuate and deepen the earthiness of the chocolate-hazelnut spread.

3 large eggs
1 cup chocolate-hazelnut spread
Pinch of salt
¾ cup almond flour
2 teaspoons porcini powder
Powdered sugar, for dusting

Preheat the oven to 350°F. Grease an 8-inch cake tin and line the bottom with a parchment paper round.

Whisk the eggs in a medium bowl until mixed. Add the chocolate spread, salt, almond flour, and porcini powder and whisk until smooth. Pour the batter into the prepared pan and bake for 30 to 35 minutes, until just set in the center. A toothpick inserted into the center should come out just a little wet.

Remove from the oven and let cool for a few minutes in the pan. Invert the cake onto a serving dish and lift off the pan. Peel off the parchment paper. When cool, dust with powdered sugar and serve. Cover any leftover torte with plastic wrap and store in the refrigerator for up to 3 days.

**SERVES 6 TO 8**

## Main Subtypes:

Almond, cashew, hazelnut, peanut, pecan, pine nut, pistachio, walnut

## Best Pairings:

Tropical fruit, honey, chocolate, roasted meat, caramel, popcorn, vinegar

## Surprising Pairings:

Citrus, shrimp, mango, pineapple, tomato

## Substitutes:

Any nut or seed may be substituted for another

Botanically speaking, most foods that we think of as nuts—among them almonds, cashews, pecans, peanuts, pine nuts, pistachios, and walnuts—are actually seeds. Both seeds and nuts should be toasted to enhance their flavor: Most of a nut's flavor is contained in its natural oils; as they heat up, they become volatile and hence more fragrant. Toasting also creates Maillard reactions, which play a large part in nuts' flavor. When improperly stored, oils in nuts can decompose and create rancid flavors, so nuts should be stored in cool, dry, dark places to minimize oxidation.

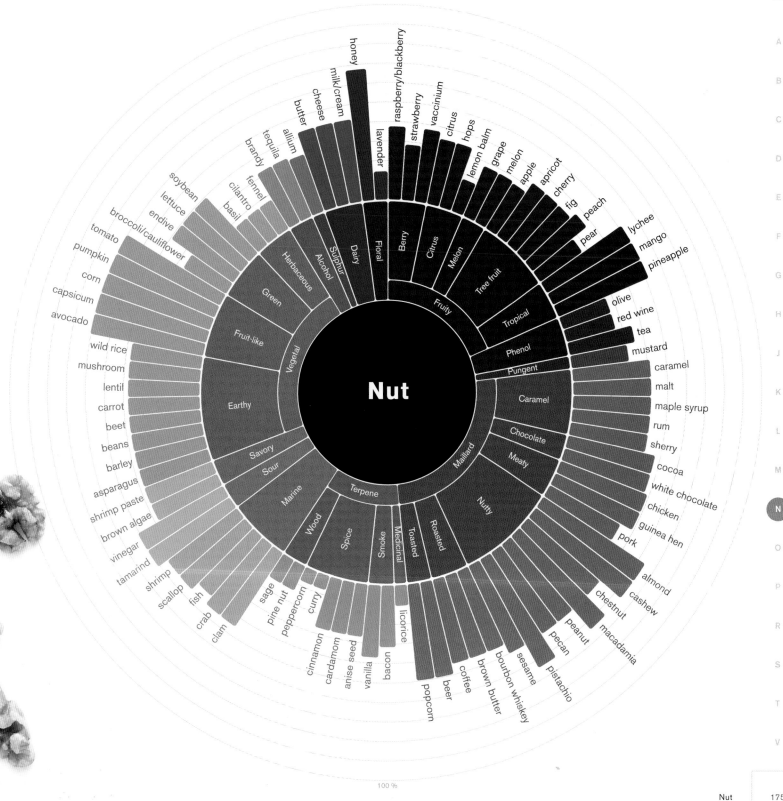

Nut

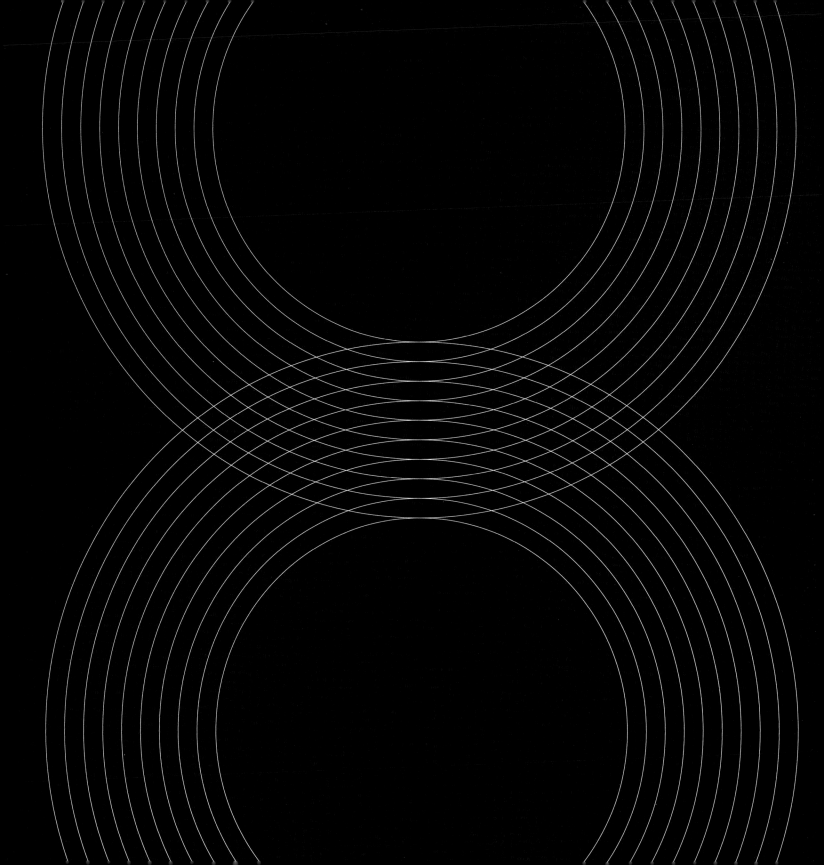

# Peanut and Lime Macarons
# with Dulce de Leche

Lime and peanuts is a fairly common combination in the cuisines of Southeast Asia, but finding these two ingredients in a cookie with dulce de leche is a surprising treat. What's more, this isn't just any cookie—it's a macaron, as delectable as it is delicate. Break out your scale and make sure you measure these ingredients precisely. (And don't be discouraged if they don't come out perfect on the first try! Great macarons take practice.)

315 grams (11 ounces) powdered sugar
165 grams (6 ounces) finely ground peanuts
Pinch of salt
1 gram (½ teaspoon) freshly ground black pepper
1 gram (½ teaspoon) finely grated lime zest
115 grams (4 ounces) egg whites
About ½ cup dulce de leche

Preheat the oven to 300°F. Line two baking sheets with silicone baking mats.

Combine half of the sugar with the peanuts, salt, pepper, and lime zest in a food processor and process for 5 seconds. Transfer to a bowl.

Beat the egg whites with a little of the remaining sugar to soft peaks with an electric mixer on low speed. Add the rest of the sugar and beat on high speed for 1 minute, until stiff peaks form.

In a few additions, sift the dry mixture over the meringue and fold it in with a spatula. Transfer the batter to a pastry bag fitted with a medium plain tip and pipe out 40 disks, each 2 inches in diameter, on the prepared baking sheets. Smooth the top of each with an offset spatula. Let the cookies sit for 15 minutes to 1 hour, until the tops are no longer sticky.

Bake for about 15 minutes, until the cookies have risen and feel light and hollow. Cool to room temperature on the baking sheets.

To assemble the cookie sandwiches, spread a layer of about 1½ teaspoons dulce de leche on the bottom (flat side) of half the cookies. Top with a second cookie and press very lightly so they stick together. Serve immediately. Store any leftover cookies in an airtight container in the refrigerator for up to 2 days.

**MAKES 20 COOKIE SANDWICHES**

Olive trees belong to the botanical family that includes jasmine and lilac. The fruit of the olive tree has a harsh, bitter taste, and must go through a process of extensive rinsing, curing, and fermentation to mellow out the phenolic compounds that make them so unpalatable. The type of curing and fermentation the olives undergo will greatly influence their final flavor: Brine-cured olives have a more mellow, well-rounded flavor, while oil- and dry-cured olives have a more intense, concentrated flavor. Olives are usually prized for their salty, fatty taste, but their actual flavor comes from strong floral aromas (olives share 69 percent of their compounds with lavender) and fruit aromas (72 percent match for nectarines). Olive and fruit pairings are doubly desirable as a salty-sweet combination.

**Best Pairings:**

Citrus, peach, plum, almond, honey, roasted meat, yogurt, cream

**Surprising Pairings:**

Chocolate, star anise, peach, nectarine, lemongrass

**Substitutes:**

Black garlic, fermented black beans, preserved lemon, capers

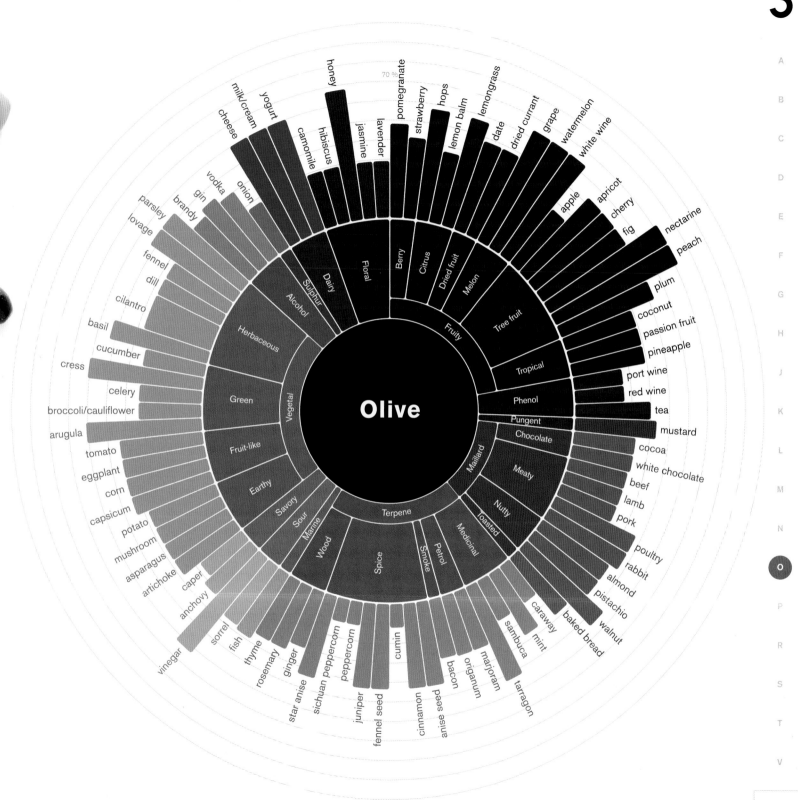

Olive

# Lemon Curd with Crunchy Olives

Olives and citrus have 60 percent of their aromatic compounds in common, so it's no surprise that they often find themselves in savory settings like martinis, hors d'oeuvres, and tapenades. Here we thrust this classic combination into a new setting: dessert. In this dish, olives are sweetened with honey, then dehydrated in the oven to intensify their flavor and create a crispy texture. This mixture is set atop a traditional lemon curd, which is finished with extra-virgin olive oil rather than the typical butter to double down on the olive flavor.

2 teaspoons honey
1 teaspoon olive brine
1 cup pitted niçoise olives
4 large egg yolks
2 teaspoons cornstarch
Grated zest of 2 lemons
½ cup fresh lemon juice
¾ cup sugar
¼ cup extra-virgin olive oil

GARNISH (OPTIONAL)

Chopped toasted almonds
Shortbread cookies

Preheat the oven to 200°F. Line a rimmed baking sheet with parchment paper or aluminum foil.

Stir together the honey and olive brine in a small bowl until smooth. Add the olives and toss well to coat. Spread the olives on the prepared baking sheet in a single layer. Bake until crisp, about 50 minutes. Remove from the oven and let cool to room temperature on the baking sheet. Store in an airtight container at room temperature for up to 5 days.

In a medium saucepot, whisk together the egg yolks, cornstarch, lemon zest, lemon juice, and sugar. When smooth, place the pot over medium heat. Stir constantly, scraping the bottom with a wooden spoon to prevent scorching. Cook until the mixture thickens and begins to bubble around the edge. Remove from the heat and whisk for 1 minute to let the heat dissipate.

Transfer the curd to a bowl and beat with an electric mixer until cooled to room temperature. (To speed this process, you can set the bowl over a bowl of ice.) When the curd is cool, slowly pour in the olive oil while continuing to beat on low speed. Beat until the curd is smooth. Transfer the curd to smaller container, press plastic wrap onto the surface to prevent a skin from forming, and refrigerate until you're ready to serve. The curd will keep in the refrigerator for up to 1 week.

To serve, divide the lemon curd among four dishes and top with dehydrated olives. If you like, sprinkle with toasted almonds and crumbled shortbread cookies.

**SERVES 4**

*Pisum sativum*—which we know simply as pea plants—includes small, round garden peas as well as sugar snap, snow peas, and field peas (aka southern peas). While we usually discard the fibrous shells of garden peas and field peas, eating only the tender seeds, we eat the entire pods of sugar snap peas and snow peas. Pea flavor is dominated by green aromas with notes of fat and nut. Garden peas are one of just a few culinary products that are often best when bought frozen. Fresh garden peas have a short eating season of late spring to summer, and they must be consumed within days of harvest for optimal taste. After harvest, peas begin to convert stored sugars into starch—meaning that within a couple of days of picking, tender sweet peas become dry and starchy. This is why most peas are frozen immediately after harvest: to preserve their tender, sweet state.

**Main Subtypes:**

Garden pea, snow pea, sugar snap pea, field pea

**Best Pairings:**

Toasted nuts, mint, alliums, fish, mushroom, berry, melon

**Surprising Pairings:**

Strawberry, coconut, coffee, apple

**Substitutes:**

Haricots verts, fava beans, edamame

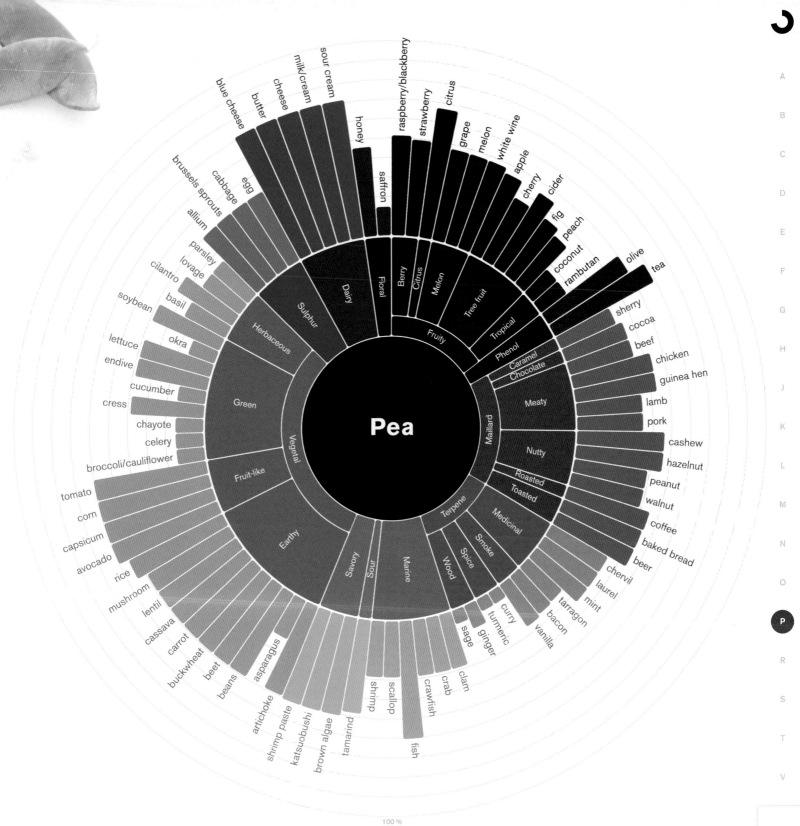

# Sweet Pea, Pork, and Coconut Tacos

The inspiration for this recipe was the too-intriguing-to-ignore pairing of peas, pork, and coconut. Plus, a taco! Begin with bright pea purée, add coconut sambal (a typical condiment from Sri Lanka that is good on everything), and finish with juicy slices of garlicky pork.

## PORK

6 tablespoons soy sauce
¼ cup packed light brown sugar
½ head garlic (cut crosswise)
2-inch piece fresh ginger, thinly sliced
2 pounds pork belly or boneless pork shoulder

## SWEET PEA PURÉE

2 tablespoons unsalted butter
6 scallions, thinly sliced
½ cup unsweetened coconut milk
2½ cups garden peas (shelled fresh or thawed frozen)
1 tablespoon chopped fresh mint

## TACOS

12 corn tortillas
Coconut Sambal (recipe follows)
¼ cup grated queso fresco

*Marinate the pork:* Combine the soy sauce, sugar, garlic, and ginger in a bowl. Add 1 cup hot water and stir to dissolve. Add 2 cups ice cubes to chill. Place the pork in a heavy-duty zip-top plastic bag, add the brine, and seal. Refrigerate overnight.

Preheat the oven to 400°F. Remove the pork from the marinade and pat dry. Place the pork in a roasting pan with a rack and cover tightly with aluminum foil. Roast for 2½ hours, until tender. The meat should be easily pierced with a fork and pull apart with little effort.

Remove the pan from the oven and let sit for 30 minutes before uncovering. If not serving immediately, wrap tightly and refrigerate for up to 7 days.

*Make the pea purée:* Melt the butter in a small sauté pan over medium heat. Add the scallions and sauté until tender. Add a tablespoon or two water if needed to prevent browning. Stir in the coconut milk and reduce by half. Combine the peas, scallion mixture, and mint in a blender and puree. Keep warm until ready to serve.

*Assemble the tacos:* Heat a broiler, grill, or cast-iron pan. Quickly sear the pork on all sides. Thinly slice. Warm the tortillas. Spread some pea purée on each tortilla. Top each with slices of pork, then sprinkle with coconut sambal and 1 teaspoon queso fresco. Serve immediately.

## COCONUT SAMBAL

1 jalapeño chile, seeded and chopped
½ cup coarsely chopped fresh cilantro
1 teaspoon ground turmeric
Juice of 1 lime
½ teaspoon light brown sugar
1 cup grated fresh coconut or thawed, drained frozen coconut

Combine all ingredients in a food processor. Pulse until finely chopped. Best when fresh, or store in an airtight container in the refrigerator for up to 5 days.

**Makes 1½ cups**

**SERVES 4**

Botanically, apple, pear, and quince are all pome fruits, which contain a hard-shelled core of seeds surrounded by edible flesh. This family of fruit, whose Latin name used to be Pomoideae (now Maloideae), is a subfamily of the rose family—so it is no surprise that floral aromas play an important role in their respective flavors. Vegetal/green and other fruit aromas also contribute to the flavor of apples, pears, and quince. Apples can be selectively bred for specific traits and uses. There are over 7,500 different cultivars raised to be consumed raw, cooked, or pressed into juice and cider. Pears tend to have a softer flesh than apples when ripe, but also contain stone cells that can give the flesh a gritty texture. These stone cells result from the formation of lignins within the cells of the flesh of pears and quince. Otherwise rarely found in foods, lignins help form the firm structure of wood and bark. Quince, unlike apple and pear, is rarely eaten raw, because most varieties have hard, astringent flesh. Quince has a strong perfume and can be added to cooked dishes that include apples or pears to increase the flavor of those ingredients. Quince is also naturally high in pectin, and thus makes excellent jellies and jams.

**Main Subtypes:**

Apple, pear, quince

**Best Pairings:**

Citrus, seafood, wine, brandy, bourbon, vinegar, olive, olive oil, spice, dairy, nuts

**Surprising Pairings:**

Basil, crab, sage, olive, peas

**Substitutes:**

Apples and pears may be substituted for one another; also: green tomato, persimmon, or quince (in cooked applications)

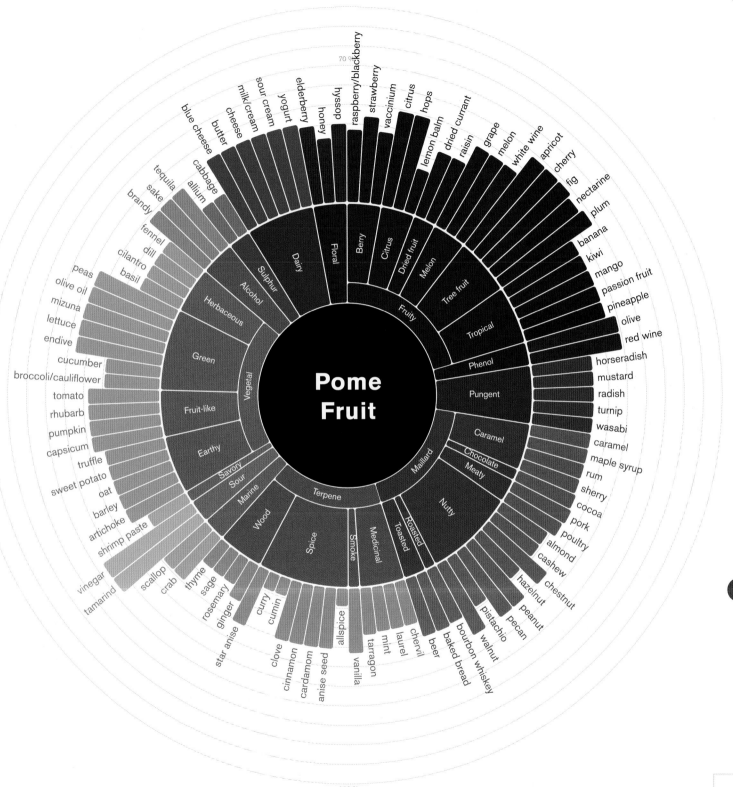

Pome Fruit

# Green Apple, Basil, and Strawberry Salsa

You've probably enjoyed the combination of apples and strawberries in many a fruit salad. But when have you seen basil and tomatoes in the mix? And when have you tasted all of these flavors with roasted fish or poultry? Apples and basil make for a particularly good pairing on their own, but on top of fish these ingredients' flavors transform into something truly unique and delicious. Serve with grilled or roasted fish or poultry, or as a bruschetta on toasted bread atop goat cheese, Parmesan, or pecorino.

1 green apple (such as Granny Smith), cored and diced
1 cup cherry tomatoes, cut in quarters
1 cup hulled and quartered strawberries
Kosher salt and freshly ground black pepper
1 tablespoon lemon juice, apple cider vinegar, tarragon vinegar, or wine vinegar
3 tablespoons extra-virgin olive oil
1 tablespoon chopped fresh basil

Combine the apple, tomatoes, and strawberries in a bowl. Season to taste with salt and pepper and toss gently to combine. Add the lemon juice, olive oil, and basil and mix again. Serve immediately; this salsa does not keep well.

**MAKES ABOUT 3 CUPS**

Pomegranate is an ancient fruit that is typically consumed for its seeds or juice, which can be reduced into molasses by slowly cooking it until it reaches a thick, syrupy consistency. Tannins can give both pomegranate seeds and juice a bitter, astringent taste, especially if the fruit is underripe and has not yet developed natural sugars, but pomegranate molasses has an intense sweet-sour taste. Pomegranate flavor is mostly fruity, with aromas of citrus, melon, and tree fruit, although it's also characterized by sharper aromas such as anise and mint.

**Best Pairings:**

Citrus, melon, basil, tea, lemongrass, chiles, avocado

**Surprising Pairings:**

Mushroom, basil

**Substitutes:**

Cranberries, raspberries, blood orange

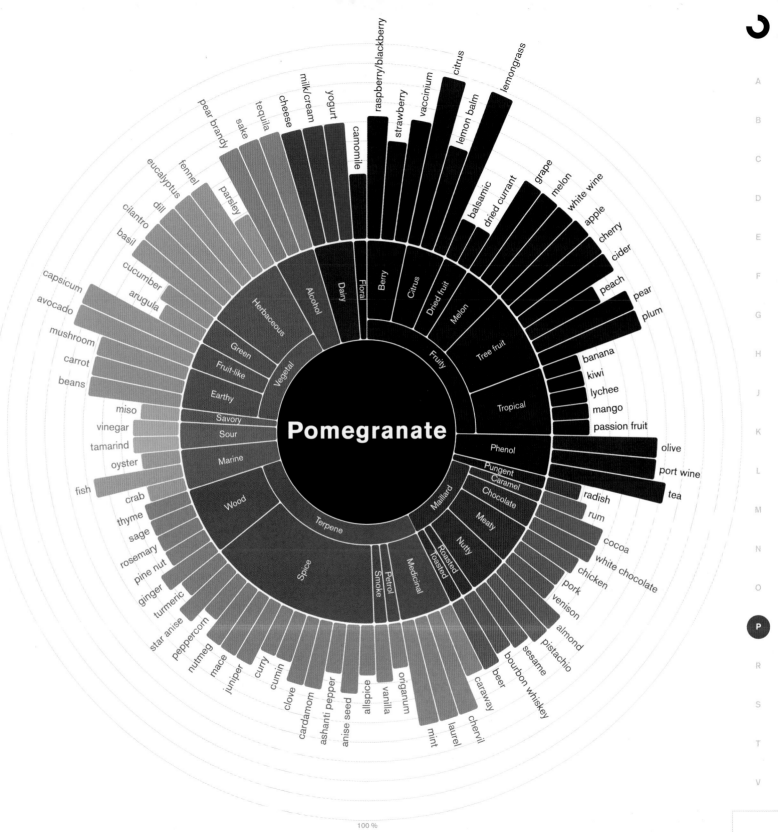

Pomegranate

# Spicy Yogurt and Lemongrass–Marinated Chicken with Pomegranate

Besides being gorgeous, this dish is incredibly complex. The pomegranate, yogurt, and herbs play up each other's hidden floral notes, while the lactones in the chicken blend seamlessly with fruity lactones and esters in the pomegranate. If you're feeling bold, try substituting duck breast or boneless quail for the chicken in this recipe.

1 tablespoon minced garlic
Grated zest and juice of 1 lime
½ cup plain yogurt
2 tablespoons minced fresh cilantro stems
1 tablespoon chopped fresh mint
1 jalapeño chile, minced
One 3- to 4-inch stalk lemongrass, thinly sliced,
    or several dashes lemongrass bitters (page 157)
Kosher salt
3 pounds boneless, skinless chicken thighs
Olive oil

### GARNISH

Sliced avocado
Fresh cilantro leaves
Fresh mint leaves
Pomegranate seeds

In a large bowl, combine the garlic, lime zest and juice, yogurt, cilantro stems, mint, jalapeño, and lemongrass. Mix well and season to taste with salt. Reserve ¼ cup of the yogurt mixture.

Add the chicken to the remaining yogurt and mix to coat evenly. Cover and refrigerate until you're ready to cook, at least 2 hours and up to 24 hours.

Heat a grill until medium-hot and oil the grates well. Set a wire rack over a baking sheet and place it next to the grill. Remove the chicken from the marinade and wipe off any excess. Discard the marinade. Lightly coat the chicken with oil and season with salt. Grill the chicken over the heat, flipping it once, about 12 minutes per side, until a thermometer inserted into the thickest part of the meat registers 160°F on an instant-read thermometer. Transfer the chicken to the wire rack to rest for 3 to 5 minutes before serving; it should reach 165°F.

Serve the chicken garnished with the reserved yogurt, avocado, cilantro, mint, and pomegranate seeds.

Pork, the flesh of domesticated pigs, is the most commonly consumed meat in the world. It is consumed both fresh and in a variety of preserved forms—including bacon, ham, sausages, terrines, and salt pork—that are collectively known as charcuterie. Pork's flavor is largely determined by the breed of the pig and the conditions under which it is raised. Beginning in the mid-1950s, consumer demand led to a trend toward leanness in pork, but lower fat content also made the meat less flavorful. In recent years, pork fat—and the breeds that provide it—has made a comeback. Although they are higher-priced than conventional pork, heritage breeds yield much more flavorful meat with a higher ratio of fat to lean and a richer coloring that is more red than white. Like grass-fed beef, pasture-raised pigs develop high quantities of omega-3 fatty acids and linoleic acid. Maillard flavors make the best pairings for pork, especially darker aromas like caramel or roasted; pork also goes very well with the fruity aromas found in tropical fruit, chiles, and tomatoes.

**Best Pairings:**

Bourbon, brown butter, Jerusalem artichoke, sweet potato

**Surprising Pairings:**

Clam, coffee, popcorn

**Substitutes:**

Poultry, game meat

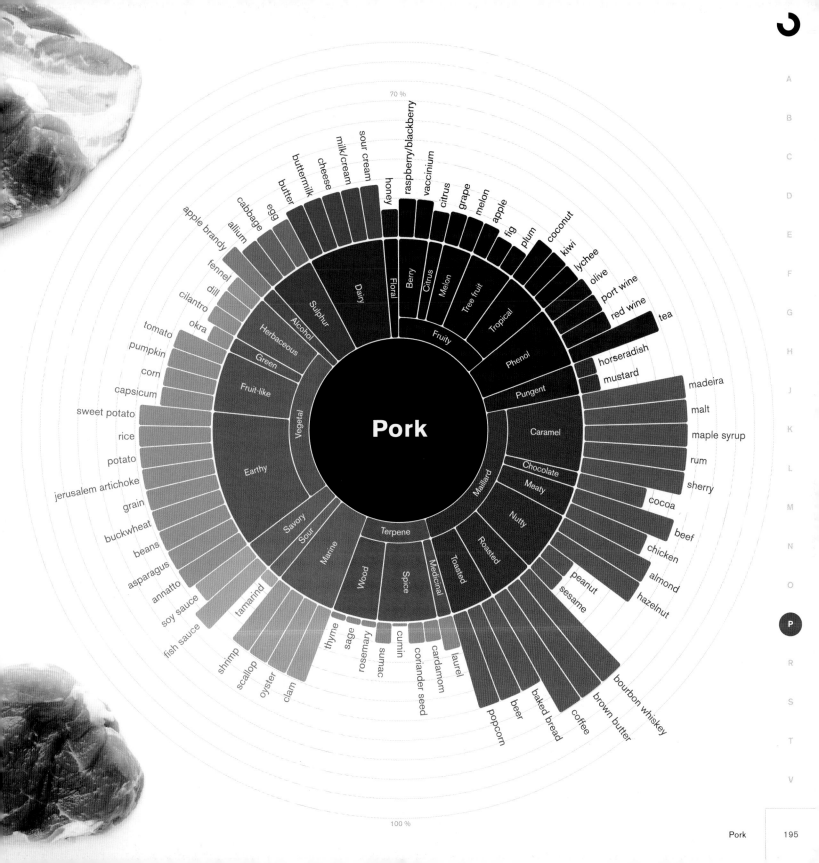

Pork

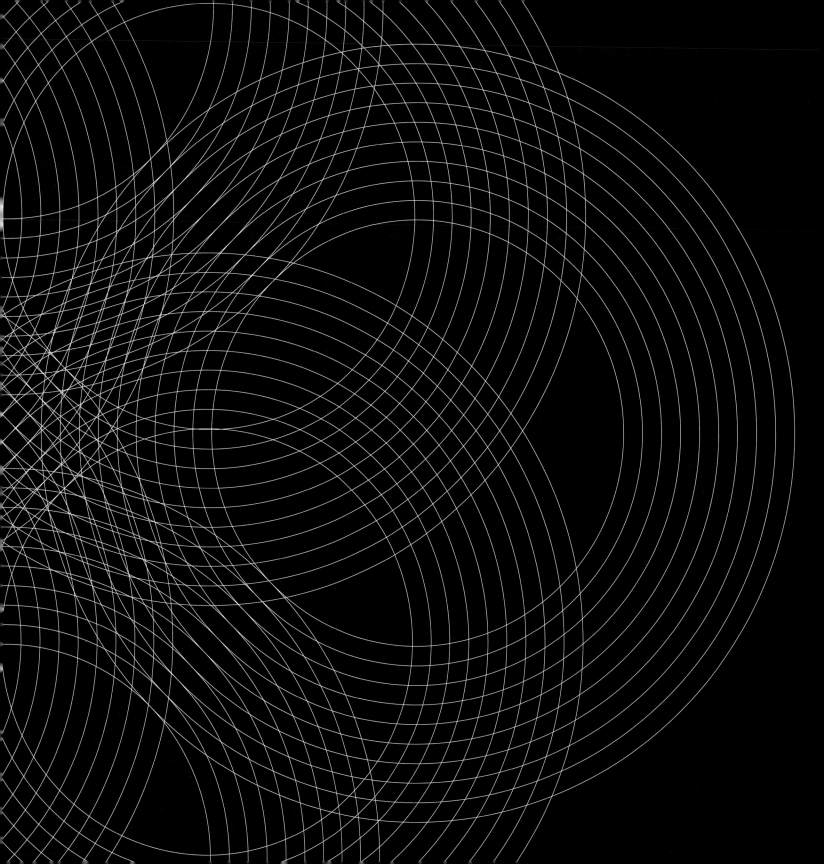

# Blueberry and Horseradish Jam

The idea for this recipe came to us when trying to look beyond the direct connections between ingredients. Follow us for a moment here: Pork pairs very well with fruit aromas, since lactones—which have an intensely fruity aroma—are major contributors to pork flavor. One particular fruit, blueberries, has latent wood and pine aromas. We kicked up those pine-y notes with horseradish and put these three flavor profiles together in a jam with a spicy, savory edge. This stuff is absolutely delicious spread on toast with bacon for breakfast, and it is also amazing on roast pork sandwiches, or simply served as a condiment with pork loin.

1 cup dried blueberries

1 teaspoon kosher salt

½ teaspoon freshly ground black pepper

2 cups apple juice

4 pods cardamom

Zest of 1 lemon, removed in strips with
    a vegetable peeler

Zest of 1 orange, removed in strips with
    a vegetable peeler

1 pound blue berries, fresh or frozen (about 2½ cups)

3 tablespoons low- or no-sugar pectin

¼ cup grated fresh horseradish

Combine the dried blueberries, salt, pepper, and apple juice in a medium saucepot and bring to a simmer over medium-low heat. Tie the cardamom, lemon zest, and orange zest in a piece of cheesecloth and add to the pot. Continue to simmer until only ¼ cup of the juice remains.

Add the frozen berries and stir well. Cook over medium-low heat until the mixture comes to a simmer. Cook until thickened, about 15 minutes, stirring occasionally to prevent scorching.

Remove from the heat. Remove and discard the cheesecloth with the cardamom and zests. Sprinkle the pectin over the surface of the berries and let stand for 1 minute, then stir until completely dissolved. Stir in the horseradish. Serve immediately. Or transfer to a clean glass jar, seal tightly, and store in the refrigerator for up to 3 months.

**MAKES APPROXIMATELY 2½ CUPS**

All cultivars of modern potatoes can be traced back to a single origin in southern Peru. After the ships of Spanish explorers returned from South America in the sixteenth century, this simple plant became a staple across much of the globe. Potatoes have a very mild, somewhat bland flavor, though the aroma of a baked potato is actually quite complex, being comprised of 228 different aromatic compounds. Much of a potato's flavor is generated in the cooking process, meaning Maillard aromas play an important role. Milk, butter, sour cream, or cheese are the most classic pairing for potatoes—and data shows them to be the best chemical match, as well. Earthy and fruit-like vegetal aromas are also main contributors to potato flavor; unsurprisingly, the other ingredients in which these fruit-like vegetal aromas are often found—tomato, eggplant, and capsicum—are nightshades, the botanical family to which potato also belongs. (Not included here are sweet potatoes, which have flavors more akin to winter squash and thus can be found on page 222.)

**Best Pairings:**

Butter, cheese, sour cream, tomato, capsicums, corn, toasted nuts, breadcrumbs

**Surprising Pairings:**

Pistachio, avocado, sake

**Substitutes:**

Zucchini, chayote, cauliflower, Jerusalem artichoke, parsnips

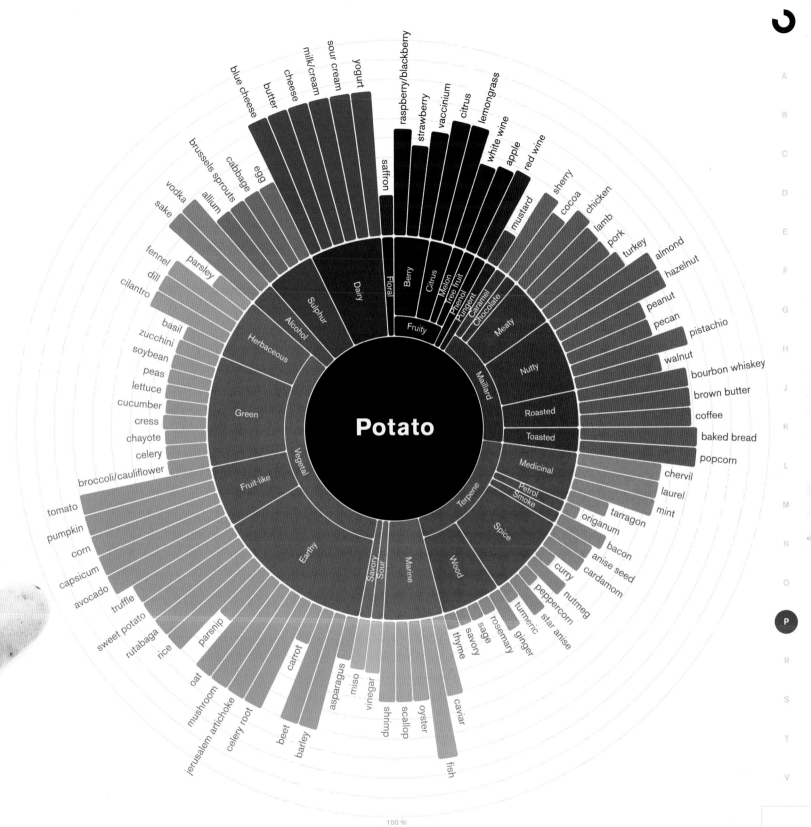

Potato

# Beet and Blue Cheese Rösti Potatoes with Whipped Avocado

Earthy and fruit-like flavors are the most prominent ones in potato, and therefore are the best pairing for the humble spud. Beets are loaded with compounds that provide both types of aromas. In this flavor- and color-packed take on the classic rösti potatoes, blue cheese punctuates these two ingredients' earthy flavors, while avocado lends creamy richness to the dish.

2 pounds small Yukon gold potatoes
Kosher salt
2 tablespoons unsalted butter
4 scallions (green and white parts), thinly sliced
1 tablespoon olive oil, plus more if needed
1 cup grated peeled raw red or golden beets
¼ cup crumbled blue cheese
Whipped Avocado (recipe follows)

Pierce the potatoes with a fork and place them in a medium saucepot. Add cold water to cover by 3 inches and season with salt. Bring to a boil, then reduce the heat to a simmer and cover the pot. Cook until the potatoes are completely tender, about 20 minutes. Drain the potatoes and let them to cool completely; ideally, refrigerate them for 2 to 3 hours. This may be done up to 3 days in advance; store the potatoes in the refrigerator.

Set a wire rack over a baking sheet and place it next to the stove. Peel the potatoes with a paring knife. Grate the potatoes on the large holes of a box grater. Melt the butter in a medium cast-iron or nonstick skillet over medium heat. Add the scallions and sauté until tender, about 2 minutes. Stir in the grated potato and mix well. When the potato begins to sizzle, stir in the olive oil, beets, and blue cheese. Press the mixture into a firm cake and let cook undisturbed until golden brown and crisp on the bottom, about 5 minutes. With a spatula, carefully slip the cake onto a plate, then invert it back into the pan to brown the other side. Add more oil if needed. When both sides are golden brown, slide the cake onto the rack to cool slightly.

Slice the cake into 6 to 8 wedges. Serve topped with whipped avocado.

## WHIPPED AVOCADO

1 avocado (preferably Hass)
1 tablespoon fresh lime juice
¼ cup plain yogurt or sour cream
¼ teaspoon ground cumin
Kosher salt and freshly ground black pepper

Cut the avocado in half, remove the pit, and scoop out the flesh. Combine the avocado, lime juice, yogurt, and cumin in a food processor and blend until smooth. Alternatively, mash the avocado in a bowl, add the lime juice, yogurt, and cumin, and purée with an immersion blender until smooth. Season to taste with salt and pepper. Serve immediately. Or transfer to a glass jar or plastic container and place plastic wrap directly on the surface. Seal tightly and store in the refrigerator for up to 3 days.

**Makes about 1 cup**

**Main Subtypes:**

Chicken, turkey, duck, quail, squab, pheasant, guinea hen, goose

**Best Pairings:**

Butter, cheese, root vegetables, grains, dill, chervil

**Surprising Pairings:**

Crawfish, banana

**Substitutes:**

All varieties of poultry may be substituted for one another

Poultry are domesticated birds raised for eggs or meat. While we commonly think of any bird's breast meat as "light" and its leg meat (thigh and drumstick) as "dark," the truth is actually more complex. Most domesticated birds are flightless, meaning the breast muscles do very little work, allowing them to remain white and very tender. The downside is that these muscles do not develop much flavor. The muscles that make up the leg meat, on the other hand, do significantly more work; this means they have more tough connective tissue, but also that they are darker and more flavorful. The catch is that birds that have not had their flying abilities bred out of them—such as duck, pheasant, quail, pigeon (squab), and geese—actually have dark breast meat. Thus poultry can be divided into two unofficial categories: those with light breast meat and those with dark breast meat. Birds in the former category (chicken, turkey, and guinea hen) have delicately flavored breast meat; birds with dark breast meat (duck, pheasant, quail, pigeon/squab, and geese) carry aromas of fat, nuts, and faint barnyard in the meat of both the breast and leg.

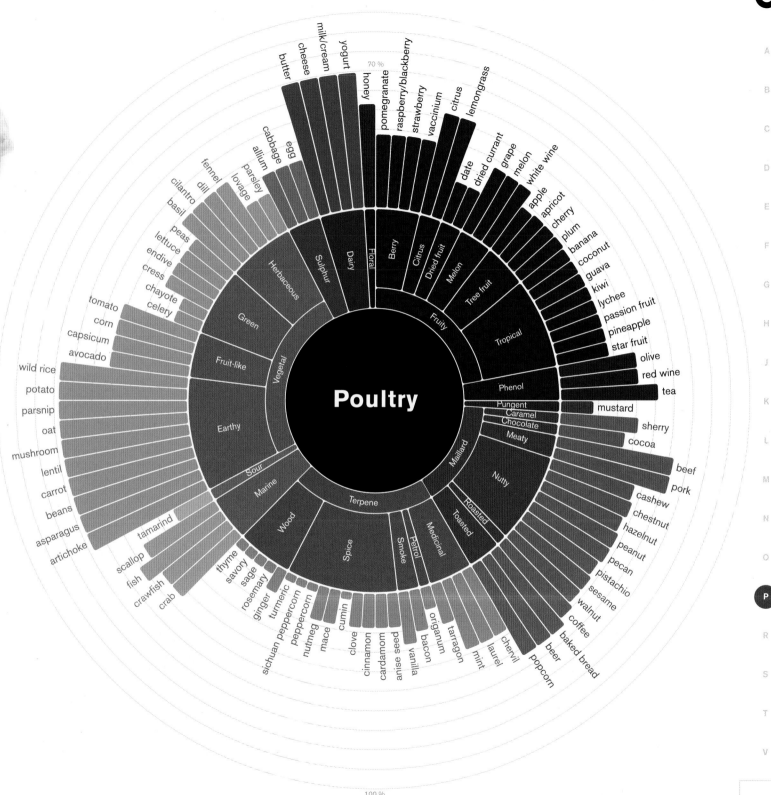

**Poultry**

# Apricot, Sage, and Pistachio Chutney

This chutney is extremely versatile and can be used as either a condiment or a stuffing for just about any roasted poultry, from turkey to chicken to duck. To employ it as a stuffing, spread on boneless cuts of breast or leg and roll up before roasting or grilling. While the flavors here are particularly well suited for poultry, it also pairs beautifully with grilled or roasted beef and lamb.

2 cups diced dried apricots
2 cups dry white wine
¼ cup white wine vinegar
3 branches fresh sage
1 teaspoon fennel seeds, crushed
2 tablespoons olive oil
Finely grated zest and juice of ½ orange
¼ cup chopped toasted pistachios or pecans
1 tablespoon minced jalapeño chile
Kosher salt and freshly ground black pepper

Combine the apricots, wine, vinegar, sage, and fennel seeds in a small saucepot. Bring to a boil, reduce the heat to a simmer, and cover the pot. Cook for 5 minutes, until the apricots are plumped, then remove from the heat and let stand for 15 minutes.

With a slotted spoon, transfer the apricots to a bowl. Discard the sage, but save the remaining liquid.

Mix the olive oil, orange zest and juice, pistachios, and jalapeño into the apricots and season to taste with salt and pepper. If the mixture is too dry add some of the reserved liquid. Use immediately or store in the refrigerator in an airtight container for up to 2 weeks.

**MAKES ABOUT 3 CUPS**

**Best Pairings:**

Chervil, basil, lettuce, peas, yogurt, sour cream, apple, citrus

**Surprising Pairings:**

Basil, passion fruit, beer

**Substitutes:**

Jicama, turnip, water chestnut

Radish is a member of the Brassicaceae family; its species, *Raphanus sativus*, is a close relative of *Brassica oleracea* and *Brassica rapa*. Radishes are commonly grown for their crunchy, spicy roots, although the plants' tops are also edible. Radish leaves have a peppery bite similar to arugula. Like other members of the Brassicaceae family, radishes derive much of their flavor from sulfur compounds. In this particular species, the most important sulfur compounds are called isothiocyanates. These nitrogen-containing molecules are responsible for the "spiciness" of raw radishes; they are also found in mustard, horseradish, and wasabi. Radishes are typically classified by their growing season; the majority reach maturity in spring, but larger varieties (such as daikon, Korean, and black radishes) are harvested in winter months.

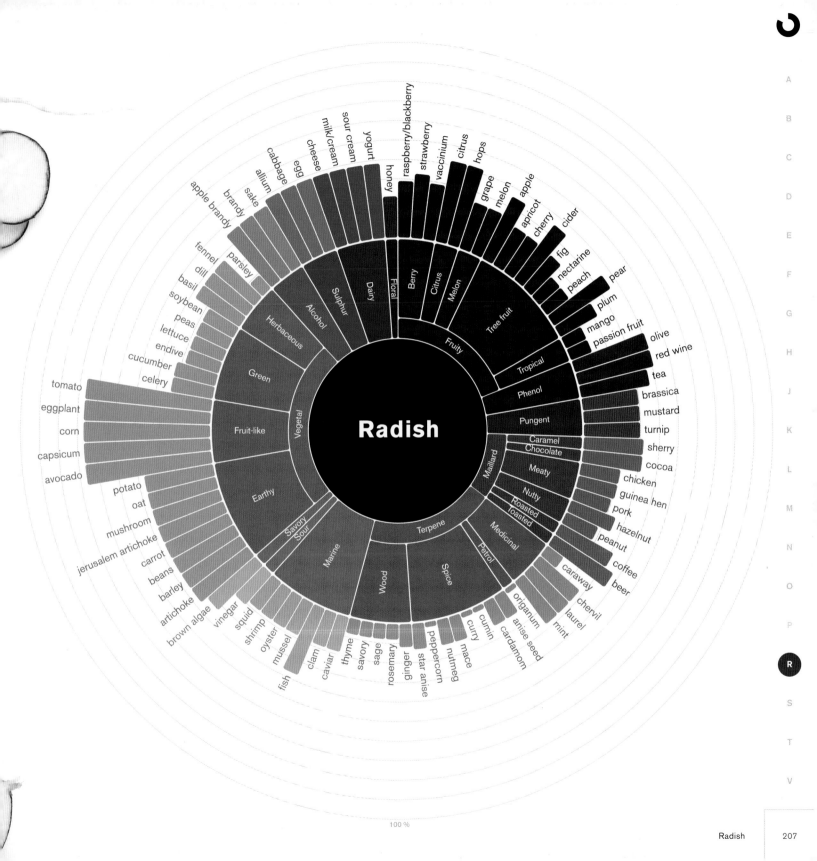

Radish

# Radish-Apple Salad
# with Basil Vinegar

Apples and basil both pair excellently with radish. Here, we make a deliciously crunchy and spicy salad out of these ingredients, and use a versatile flavor-infusing technique to make a deeply aromatic and complex dressing. Basil vinegar hits just the right notes for this recipe: It makes for some of the most flavorful radishes, whether you use it to deglaze a pan of sautéed radishes or use it as a dressing for a simple salad of sliced radishes and fresh lettuces. It also makes the perfect dressing for a tomato salad. Be sure to try out some of the variations listed below.

2 tablespoons Basil Vinegar (recipe follows)
2 tablespoons apple cider or apple juice
½ teaspoon honey or granulated sugar
1 teaspoon Dijon mustard
2 tablespoon extra-virgin olive oil
1 Granny Smith apple, cored, quartered, and thinly sliced
1 Honeycrisp apple, cored, quartered, and thinly sliced
1½ cups thinly sliced radishes (1 bunch)
1 cucumber, thinly sliced
Kosher salt and freshly ground black pepper
Fresh basil leaves, for garnish

Combine the vinegar, cider, honey, and mustard in a large bowl. Whisk to combine. Continue whisking while adding the olive oil, until the dressing is emulsified. Add the apples, radishes, and cucumber and toss well to combine. Season to taste with salt and pepper. Arrange the salad on four plates or in a large bowl and garnish with basil leaves. Serve immediately.

## BASIL VINEGAR

1 teaspoon caraway seeds
6 branches fresh basil, cut into 4 or 5 pieces each
Zest of 1 lime, removed in strips with a vegetable peeler
2 cups white wine vinegar

Combine the caraway seeds, basil, and lime zest in a glass jar. Add the vinegar and tighten the lid. Shake vigorously to promote infusion. Store in a cool, dark place for 1 week, shaking the jar occasionally.

The vinegar is ready to use after 1 week. The basil and flavorings may remain in the jar up to 1 month. After that, strain the vinegar into a clean jar, and it will keep indefinitely.

**Makes 2 cups**

## VARIATIONS

Mint: 2 teaspoons ground turmeric, 4 branches fresh mint, zest of 1 lemon, 2 cups white wine vinegar

Chile: 1 sliced jalapeño chile, 10 to 12 branches fresh cilantro, 2 cups sherry vinegar or white wine vinegar

Raspberry: 1 tablespoon black peppercorns, 1 cup fresh raspberries, 6 branches fresh thyme, 2 cups red wine vinegar

**SERVES 4**

**Best Pairings:**

Butter, cheese, citrus, berries, mushroom

**Surprise Pairings:**

Raspberry, vanilla

**Substitutes:**

Farro, barley, wheat berries, quinoa

Rice grains are the seeds of one of two particular species of grass: *Oryza sativa* (which produces what is known as Asian rice) and *Oryza glaberrima* (which produces what is called African rice). Rice is one of the oldest agricultural products in the world, with evidence of domestication dating back more than 10,000 years. Evidence also suggests that all 40,000 varieties of rice can be traced back to a single cultivar. After harvesting, the tough, outer seed covering—the bran—can either be left on the grains of rice, or scrubbed off. Brown rice, in which the bran layer has been left intact, has a firmer texture and nuttier flavor than branless white rice, whose flavor comes from a combination of floral, herbaceous, and nutty aromas. Jasmine, basmati, Thai red, and black Japonica are particularly aromatic varieties, and have more pronounced floral and nutty aromas than other types of rice.

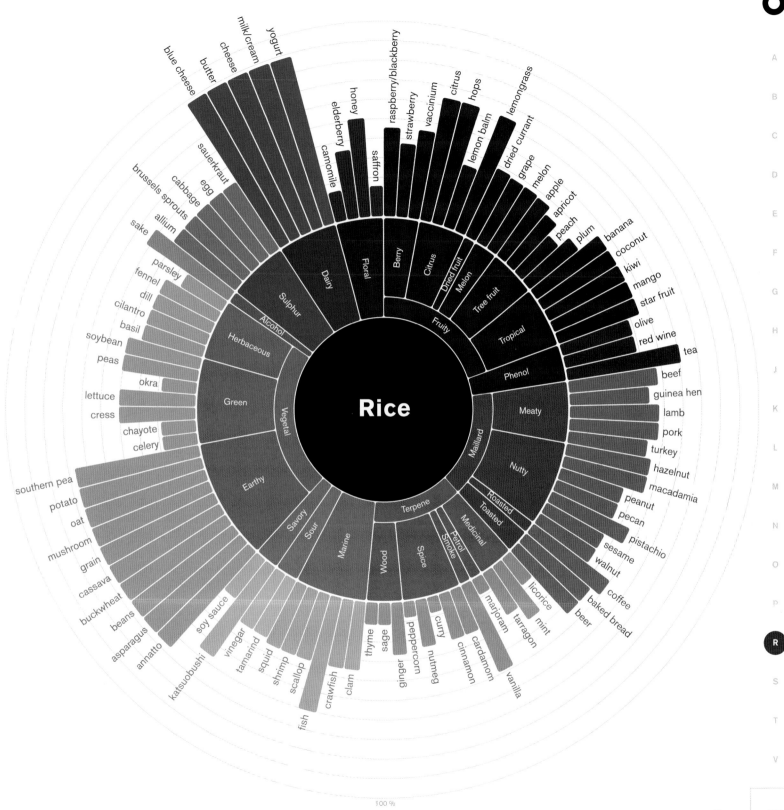

Rice

# Raspberry Rice Pudding

Rice pairs particularly well with fruit flavors, especially those from berry and citrus. In this recipe, all three ingredients come together to transform a homey dessert — rice pudding—into a refreshingly sophisticated treat.

1 tablespoon granulated sugar
2 cups fresh raspberries
½ cup Arborio rice
3 cups milk or almond milk
1 cinnamon stick
Seeds scraped from ½ split vanilla bean, or
    1 teaspoon pure vanilla extract

## GARNISH

Chopped toasted pistachios
Finely grated orange zest

Sprinkle the sugar over the raspberries and gently toss to combine. Set aside to macerate.

Combine ¼ cup sugar, the rice, milk, cinnamon stick, and vanilla seeds in a large saucepot. Bring to a gentle boil, and then turn down to a gentle simmer. Cook, stirring occasionally to keep it from sticking to the bottom, for about 30 minutes, until the rice is tender.

Mash the raspberries to a coarse purée. Strain the purée and any juices through a fine-mesh sieve into the pudding. Stir well and continue cooking until thickened, about 10 minutes more. Remove and discard the cinnamon stick.

Pour into four dessert bowls and serve immediately, or chill in the fridge. Before serving, garnish with pistachios and orange zest.

**SERVES 4**

**Main Subtypes:**

Celery root, parsnip, salsify

**Best Pairings:**

Celery, cauliflower, bourbon, coffee, lemon, basil

**Surprise Pairings:**

Vanilla, cilantro, cucumber

**Substitutes:**

Celery root, parsnip, or salsify may be substituted for one another; also: potato, Jerusalem artichoke, carrot

Technically speaking, the term "root vegetable" encompasses everything from carrots to potatoes to radishes and more. A very broad classification, it refers to the root of any plant that is used as a food source. Here we will define it more narrowly, looking at a specific group of root vegetables that have shared flavor characteristics. Celery root and parsnip are members of the same botanical family, and salsify is closely related. All three vegetables are white fleshy roots with a mild earthy, nutty, and herbaceous flavor. This class of vegetable is most commonly roasted or boiled and puréed, though celery root and parsnip may be consumed raw as well. Like all root vegetables, after these are harvested they may be stored or cellared for six months or more. Celery root (or celeriac), the knobby, bulbous root of a specific variety of celery (not common celery), should be firm and heavy. Parsnip is related to carrots and parsley, but it should not be confused with white-colored carrots. Like carrots, parsnips should be firm and crisp. Salsify, the root of a wildflower native to Europe, is also known as oyster plant because of its mild oyster-like flavor when cooked.

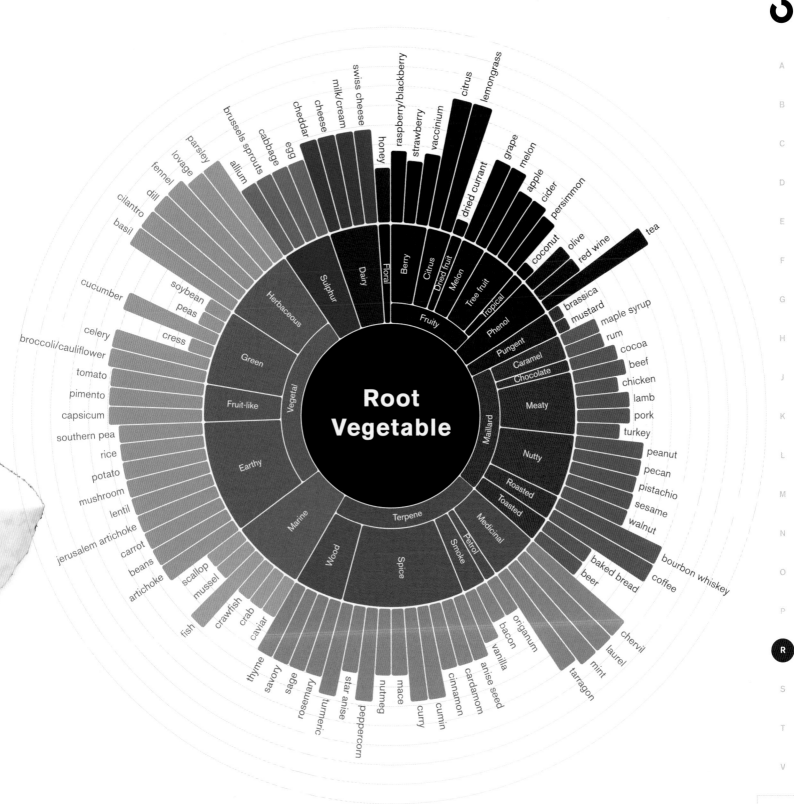

# Roasted Root Vegetables with Ginger Salsa Verde

Citrus and herbaceous are prominent aromas in parsnips, celery root, and salsify—yet many cooks overlook these subtleties when preparing and seasoning these ingredients, keeping them relatively simple. Here we take the opposite tack, playing up these flavors with a ginger salsa verde that adds a pleasant brightness to what could otherwise be a drab and familiar side dish: roasted root vegetables.

2 pounds parsnips, celery root, and/or salsify
¼ teaspoon coriander seeds, crushed
¼ teaspoon ground ginger
1 clove garlic, chopped
1 tablespoon extra-virgin olive oil
2 teaspoons kosher salt
½ teaspoon freshly ground black pepper
Ginger Salsa Verde (recipe follows)

## GARNISH (OPTIONAL)

Celery leaves
Fresh parsley
Shaved radishes

Preheat the oven to 425°F. Line a baking sheet with parchment paper.

Peel the root vegetables and cut into sections about 4 inches long. Cut the sections in half or quarters, depending on their size, so that all the pieces are about the same size.

Whisk together the coriander, ginger, garlic, olive oil, salt, and pepper in a large bowl. Add the vegetables and toss to coat. Spread the vegetables on the prepared baking sheet in a single layer and roast until easily pierced with the tip of a knife, 12 to 15 minutes. Set aside to cool slightly.

Arrange the vegetables on a platter and drizzle with the salsa verde. Top with any or all of the garnishes, as you like.

## GINGER SALSA VERDE

Leaves from 1 large bunch fresh cilantro (about 2 cups)
1 tablespoon drained capers
2 teaspoons fresh thyme leaves
1 tablespoon dried marjoram
1 clove garlic, grated
1 tablespoon grated peeled fresh ginger
½ jalapeño chile, seeded and chopped
¼ cup extra-virgin olive oil
2 teaspoons red wine vinegar
1 teaspoon grated lemon zest
Kosher salt and freshly ground black pepper

Combine the cilantro, capers, thyme, marjoram, garlic, ginger, jalapeño, olive oil, and vinegar in a food processor and blend in short bursts. The mixture should be finely chopped but not a smooth purée. Stir in the lemon zest and season to taste with salt and pepper. Serve immediately. Store any leftover sauce in a tightly sealed glass jar in the refrigerator for up to 2 weeks. Stir to recombine before using.

**Makes 2½ cups**

**SERVES 4**

**Main Subtypes:**

Zucchini, yellow squash, chayote, pattypan
squash, squash blossom

**Best Pairings:**

Asparagus, grain, mushroom, lemon, cheese,
garlic, fish sauce

**Surprise Pairings:**

Peanut, parsnip, cabbage

**Substitutes:**

Any summer squash may be substituted for
another; also: cucumber

Summer squash is the common name for a variety of squashes of the species *Cucurbita pepo*, which also includes winter squash and pumpkin. These warm-weather squashes generally have a very mild, even bland flavor, characterized mostly by vegetal aromas with faint hints of floral, nut, and fat. When selecting summer squash, look for smaller specimens, as these tend to have a better ratio of firm flesh to soft seed pockets (which tend to become mushy when cooked). Delicately flavored squash flowers, or blossoms, are a favorite early summer product; they grow from the end of all varieties of zucchini and squash and come in both male and female varieties. Once pollinated, female flowers will grow new squashes, while the male flowers grow from long stems near the base of the plant and tend to be the best for stuffing.

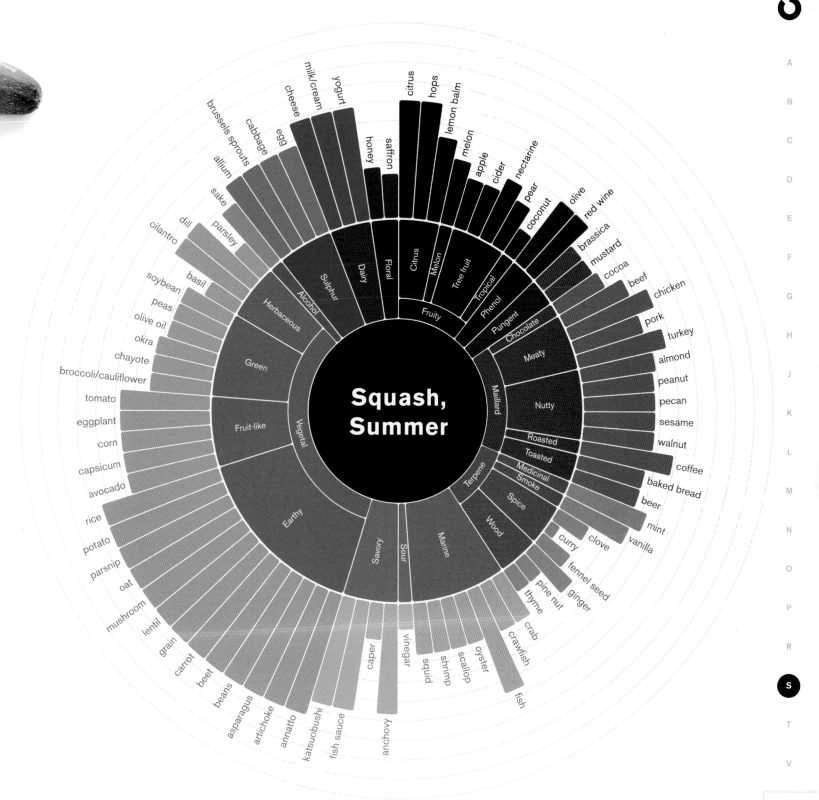

Squash, Summer

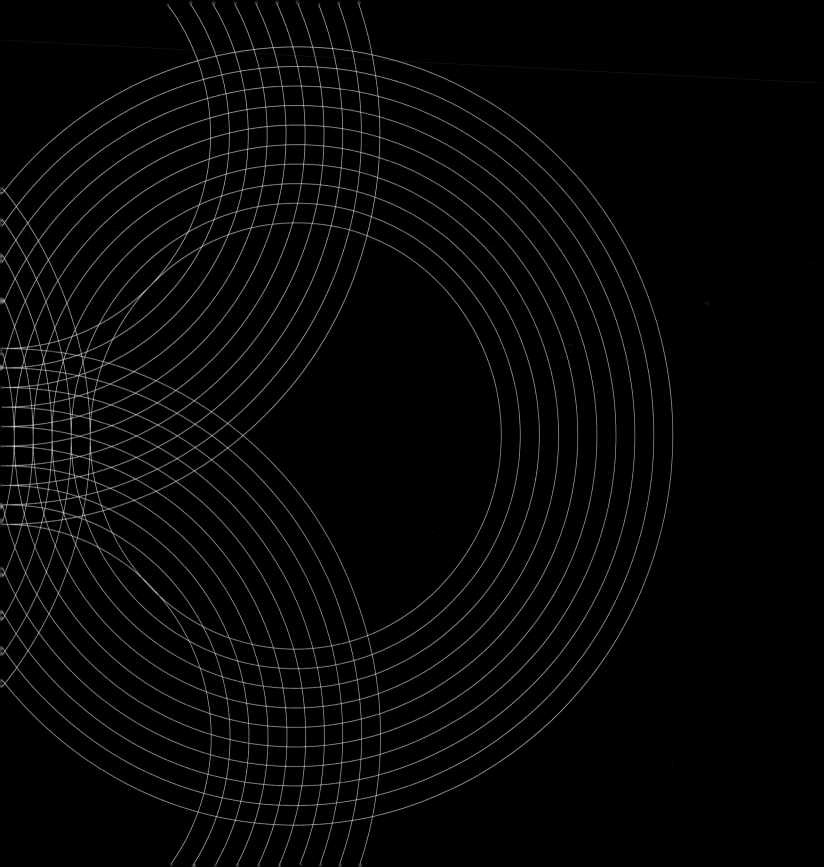

# Thai-Style Spicy Zucchini Salad

The Southeast Asian ingredients of lime, peanut, ginger, and fish sauce make for a great pairing with zucchini—which easily stands in for green papaya in this fresh take on the traditional Thai salad.

¼ cup fresh lime juice

2 tablespoons light brown sugar

1½ tablespoons fish sauce

1 tablespoon finely grated peeled fresh ginger

4 cloves garlic, minced

2 medium zucchini, trimmed and grated or julienned (about 4 cups)

½ teaspoon kosher salt

1 cup cherry tomatoes, cut in half

4 scallions (green and white parts), thinly sliced

2 tablespoons fresh cilantro leaves

1 fresh bird or Thai chile, thinly sliced

¼ cup toasted peanuts, coarsely chopped, for garnish

Whisk together the lime juice, sugar, fish sauce, ginger, and garlic in a small bowl. Set the dressing aside for the flavors to combine.

Place the zucchini in a large bowl and add the salt. Toss well to combine. Let stand 10 minutes, then transfer to a colander and squeeze dry. Break apart the squeezed zucchini and return to the bowl.

Add the tomatoes, scallions, cilantro, chile, and dressing. Toss well to combine. Adjust the seasoning to taste with salt or fish sauce and lime juice. Divide among four plates and garnish with peanuts before serving.

**SERVES 4**

There are several species of winter squash within the genus *Cucurbita*, which includes edible varieties of pumpkin and squash as well as decorative gourds. Unlike zucchini, also a *Cucurbita*, winter squash and pumpkins are rarely eaten raw, and their skin is almost always removed before they are consumed. Squash and pumpkins may grow in a wide variety of colors, shapes, and sizes. Their interiors are most commonly orange; they are bitter and starchy when raw but become tender-sweet when cooked. Despite its orange color, the flavor of winter squash is made up of strong vegetal/green aromas. The strongest pairings are with citrus and Maillard flavors. (Though not technically a winter squash, sweet potato is included here because it has a very similar flavor profile and culinary uses to those of winter squash and pumpkin.)

**Main Subtypes:**

Acorn squash, butternut squash, spaghetti squash, sweet potato, pumpkin

**Best Pairings:**

Citrus, butter, cream, cheese, caramel, toasted nut, cocoa, mushroom

**Surprise Pairings:**

Melon, clam, okra, tea

**Substitutes:**

Any variety of winter squash or pumpkin may be substituted for another

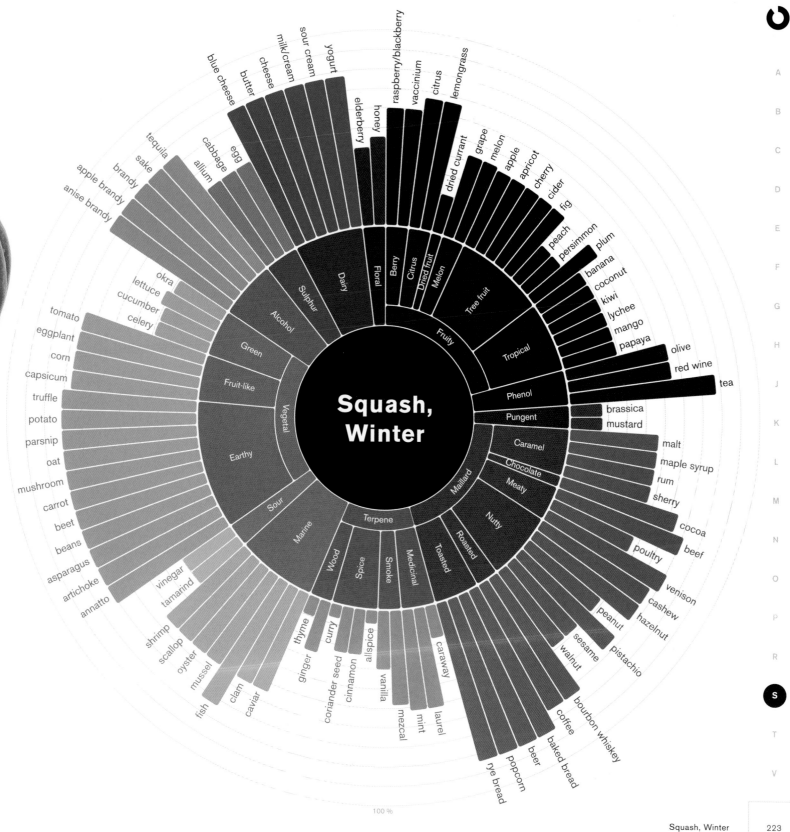

# Squash, Winter

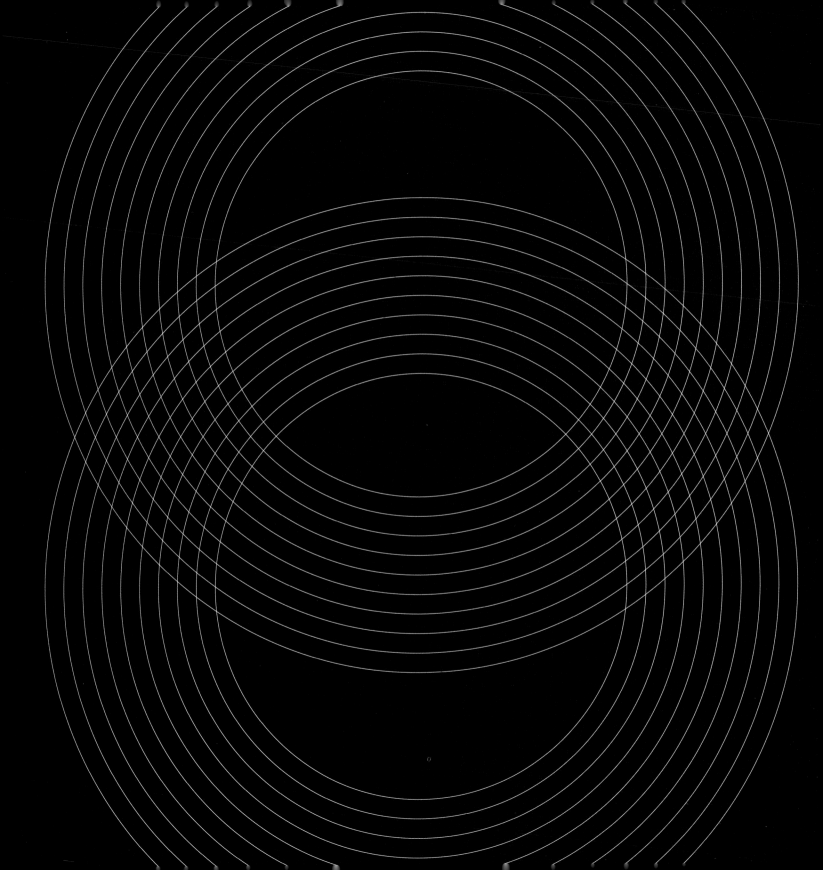

# Melon Salad with Winter-Squash Pickles

Winter squash and melons are closely related, giving them similar flavor profiles and a good pairing score. Unfortunately, one grows in spring, the other in winter—so they rarely have a chance to meet in the same dish. But we came up with a solution: Make winter-squash pickles and then add them to a July fresh melon salad. We were blown away by the results!

## WINTER-SQUASH PICKLES

¾ cup white wine vinegar or rice vinegar
¼ cup water
1 tablespoon yellow mustard seeds
1 teaspoon kosher salt
2 teaspoons sugar
½ teaspoon curry powder
2 branches fresh thyme
2 or 3 slices fresh chile (optional)
1 cup thinly sliced peeled winter squash

## MELON SALAD

½ cup plain Greek yogurt
1 teaspoon curry powder
1 teaspoon minced fresh mint leaves
Grated zest and juice of ½ lime
1 Kirby cucumber, thinly sliced (about 1 cup)
1½ cups diced peeled melon (honeydew, cantaloupe, or watermelon, or a combination)
1 cup cherry tomatoes
1 medium-hot long red chile (such as a red cowhorn), sliced
2 tablespoons olive oil
Kosher salt
2 tablespoons fresh basil leaves

*Make the pickles:* Sterilize a 1-pint glass canning jar. Combine vinegar, water, mustard seeds, salt, sugar, curry powder, thyme, and chile (if using) in a saucepot and bring to a boil. Reduce the heat to simmer and cook for 2 to 3 minutes, to develop the flavors. Add the squash and simmer until it is just tender, about 2 minutes. Transfer to the jar, cap loosely, and let cool. Seal tightly and store in the refrigerator for up to 6 months.

*Make the salad:* Combine the yogurt, curry powder, mint, and lime zest and juice in a small bowl and whisk until smooth. Set aside for the flavors to develop.

Drain the pickles, reserving 2 tablespoons of the pickling liquid; discard the thyme and chile (if using). Transfer the pickles to a large bowl. Add the cucumber, melon, tomatoes, chile, olive oil, and pickling liquid. Gently toss to combine. Season to taste with salt.

To serve, spoon about 2 tablespoons of the yogurt on each of 4 plates and spread with the back of a spoon. Mound the salad on top of the yogurt and garnish with basil leaves. Serve immediately.

**SERVES 4**

Stone fruit, or drupes, are a class of fruit that have thick flesh surrounding a pit, which in turn contains a kernel. Here we'll look specifically at drupes of the genus *Prunus,* which encompasses peaches, plums, apricots, nectarines, and cherries. All of these fruit derive their distinctive flavors from lactones, a class of aromatic compounds with intensely fruity aromas that are often used in the production of fragrances and flavorings. While all stone fruit share certain flavor characteristics, they can be divided into species with even more specific taste and flavor attributes. Peaches and nectarines are yellow- or white-fleshed fruits that are at peak ripeness from June to early September. Their flavor comes from a combination of citrus, floral, and almond aromas. Plums and apricots are green- to gold- to orange- to red-fleshed fruits that are at peak ripeness in August. Their flavor comes from fruity lactones with aromas of apricot, banana, and orange. Cherries are red-, yellow-, or white-fleshed fruits that are at peak ripeness in July. Their flavor comes from a combination of almond, fruit, and vegetal aromas.

**Main Subtypes:**

Peach, nectarine, plum, apricot, cherry

**Best Pairings:**

Cream, yogurt, basil, citrus, lemongrass, fruit brandy, corn, capsicums, wine

**Surprise Pairings:**

Beer, sage, soy sauce

**Substitutes:**

Any stone fruit may be substituted for another

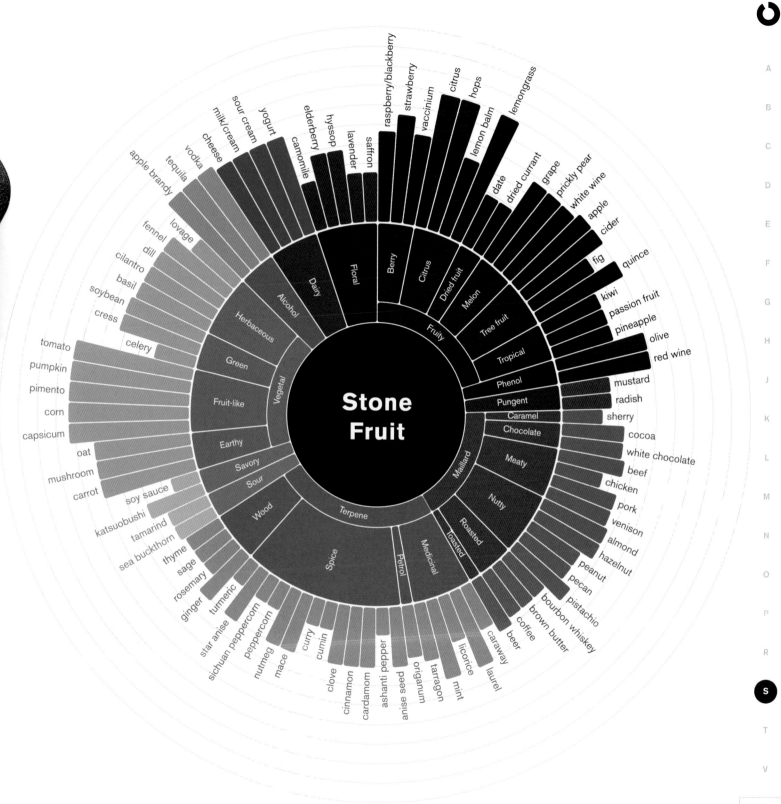

# Stone Fruit

# Pan-Roasted Pork Tenderloin with Coffee, Soy, and Peaches

Thanks to the lactones it contains, pork has latent fruit aromas that are drawn out by the peaches in this dish. Soy sauce and bourbon flip the switch on this stone fruit, giving it a savory edge. This is a ridiculously simple recipe, one that you can pull out whenever you are in a hurry but looking to impress—it packs in unexpected flavors at every step, and it can be on the table in about 20 minutes.

2 teaspoons kosher salt

1 teaspoon ground coffee

1 teaspoon ground coriander

½ teaspoon freshly ground black pepper

¼ teaspoon ground cinnamon

1 pork tenderloin (about 18 ounces), trimmed and cut crosswise into 12 medallions

3 tablespoons unsalted butter

2 tablespoons extra-virgin olive oil

2 shallots, thinly sliced

2 branches fresh sage

2 peaches, pitted and sliced about ¼ inch thick

¼ cup bourbon

2 tablespoons soy sauce

1 cup chicken stock

1 teaspoon fresh thyme leaves

Combine the salt, coffee, coriander, pepper, and cinnamon in a small bowl and mix well. Arrange the pork medallions on a plate or baking sheet and sprinkle evenly on both sides with the spices; you may not use all.

Set a wire rack over a baking sheet and place it next to the stove. Heat a large cast-iron skillet over medium-high heat. Add 1 tablespoon of the butter and 1 tablespoon of the olive oil. When the butter is melted and bubbling, cook half the pork medallions, browning them well on both sides (about 2½ minutes per side). Transfer to the rack to rest. Discard the fat from the skillet, wipe it out, and cook the rest of the medallions in the remaining oil and another tablespoon of the butter. Transfer the second batch of medallions to the rack.

Add the shallots and sage to the fat in the skillet and cook for 1 minute, then add the peach slices. Sauté the peaches until tender, 1 to 2 minutes. Remove the skillet from the heat and pour in the bourbon. Place the skillet back on the heat and reduce the liquid until nearly dry. Return the pork to the pan and add the soy sauce and chicken stock; swirl the skillet to mix the sauce. Simmer for 1 minute and stir in the remaining 1 tablespoon butter. Sprinkle with the thyme and serve immediately.

**SERVES 4**

Cane syrup is produced from freshly pressed sugar cane juice that has been boiled so water evaporates and sugar caramelizes. In this cooking process, the syrup develops deep, roasted notes. As the syrup undergoes successive boilings, it becomes darker, eventually developing the slightly sulfurous and bitter flavor characteristic of blackstrap molasses. Maple syrup is the result of a similar production process: The sap of maple trees is carefully boiled to evaporate water, but the boil is stopped before any caramelization occurs. Similar syrups are made from juiced sorghum (sorghum syrup), date palm sap (palm syrup/sugar), and coconut sap (coconut syrup/sugar). These syrups can be used as a more flavorful alternative to plain sugar or corn syrup.

**Main Subtypes:**

Cane syrup, molasses, sorghum syrup, maple syrup

**Best Pairings:**

Port wine, tamarind, sherry vinegar, apple, grain, orange

**Surprise Pairings:**

Garlic, fish, olive

**Substitutes:**

Any syrup may be substituted for another

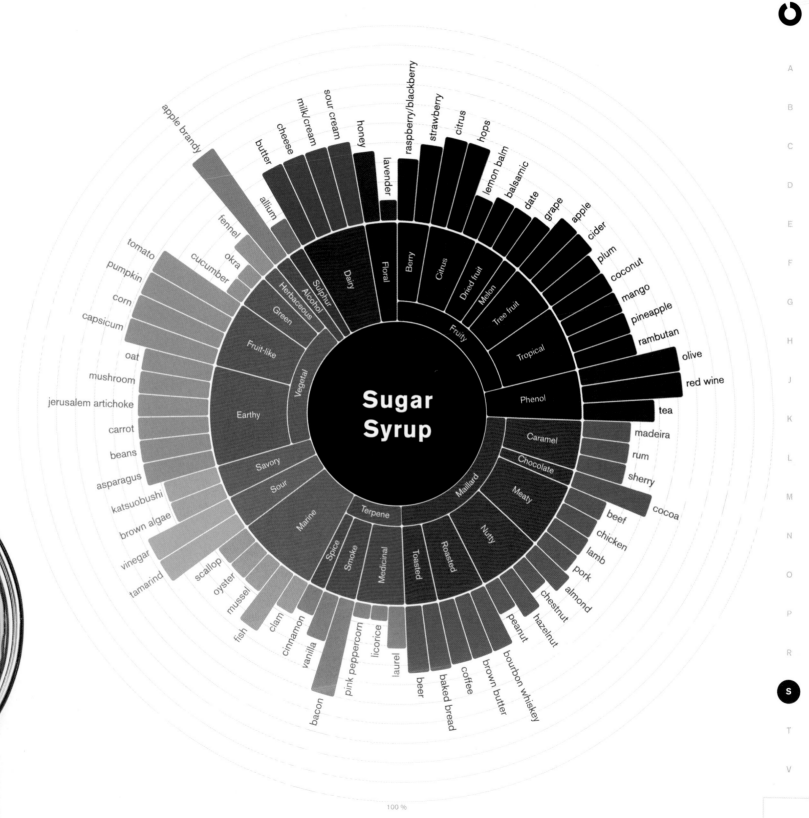

# Sugar Syrup

# Molasses-Stained Swordfish Kebabs

This molasses glaze is particularly delicious on seafood. In this recipe we've paired it with swordfish, which stands up best to grilling. But the glaze may also be brushed onto salmon before broiling, or used to marinate chicken or beef for the grill.

¼ cup sherry vinegar or red wine vinegar
¼ cup molasses or cane syrup
4 cloves garlic, minced
2 teaspoons fresh thyme leaves
Finely grated zest of ½ orange
1 teaspoon ground cumin
2 pounds swordfish, cut into 1½-inch cubes
1 red onion, cut into 1½-inch squares
1 bell pepper (any color), cut into 1½-inch squares
6 strips thick-cut bacon, cut into 1½-inch
   pieces (optional)
Eight 6-inch bamboo skewers, soaked in cold water
Kosher salt and freshly ground black pepper

Whisk together the vinegar, molasses, garlic, thyme, orange zest, and cumin in a small bowl. Divide the glaze between two containers.

Arrange the fish, onion, bell pepper, and bacon (if using) in an alternating fashion on the skewers. Fill the skewers as much as possible, leaving just ½ inch of the wood on each end. Lay out the assembled skewers on a platter or baking sheet. Season well with salt and lightly with pepper. Brush all sides of the skewers, using one container of the glaze.

Heat a grill until hot and oil the grates well. Alternatively, heat the broiler. Set a wire rack over a baking sheet and place it next to the grill or broiler. Grill or broil the skewers for 3 minutes per side, turning 3 to 4 times, until the fish is cooked medium to medium-well. Transfer the skewers to the rack to rest and immediately brush each skewer with clean molasses glaze from the second container. Serve hot with any remaining glaze.

**SERVES 4**

Tomatoes are the edible fruit of a member of the nightshade family that originated in Central and South America. Today, tomatoes are an essential part of cuisines around the world despite the fact that the plant did not arrive on the European, Asian, or African continents until the mid-1500s. Approximately 7,500 different varieties of tomato are grown worldwide. Tomatoes in general are highly aromatic, with nearly 500 different aromatic compounds coming together to create their flavor. Vegetal and fruit aromas are the most prominent, although no single compound stands out as largely responsible for tomato flavor. The taste of tomatoes is a result of the balance of natural sugars and acid present in the fruit and can range from tart to sweet; when seasoning tomatoes, you can adjust their taste by adding vinegar or sugar to fine-tune the acid/sugar balance. Tomatoes are also a natural source of umami.

**Best Pairings:**

Eggplant, corn, avocado, capsicums, peach, sesame, tamarind, lettuce

**Surprise Pairings:**

Tea, coconut, passion fruit, cardamom, berries

**Substitutes:**

Strawberries, tomatillos, persimmon, capsicums

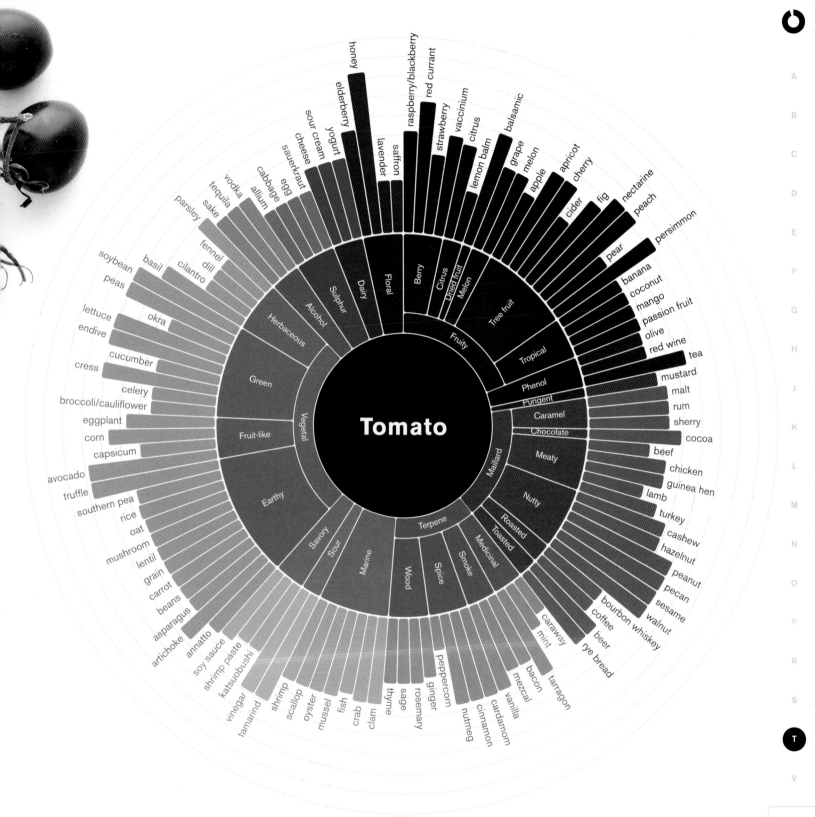

# Tomato

honey
elderberry
yogurt
sour cream
cheese
sauerkraut
egg
cabbage
allium
vodka
tequila
sake
parsley
fennel
dill
cilantro
basil
soybean
peas
lettuce
endive
okra
cress
cucumber
celery
broccoli/cauliflower
eggplant
corn
capsicum
avocado
truffle
southern pea
rice
oat
mushroom
lentil
grain
carrot
beans
asparagus
artichoke
annatto
soy sauce
shrimp paste
katsuobushi
vinegar
tamarind
shrimp
scallop
oyster
mussel
fish
crab
clam
thyme
sage
rosemary
ginger
peppercorn
nutmeg
cinnamon
cardamom
vanilla
mezcal
bacon
mint
caraway
tarragon
coffee
beer
rye bread
bourbon whiskey
walnut
pecan
sesame
peanut
hazelnut
cashew
turkey
lamb
guinea hen
chicken
beef
cocoa
sherry
rum
malt
mustard
tea
red wine
olive
passion fruit
mango
coconut
banana
persimmon
peach
nectarine
pear
fig
cider
cherry
apricot
apple
melon
grape
balsamic
lemon balm
citrus
vaccinium
strawberry
red currant
raspberry/blackberry
saffron
lavender

lavender
saffron
Floral
Berry
Citrus
Dried fruit
Melon
Dairy
Sulphur
Alcohol
Herbaceous
Green
Fruit-like
Vegetal
Earthy
Savory
Sour
Marine
Wood
Spice
Smoke
Medicinal
Toasted
Roasted
Nutty
Meaty
Chocolate
Caramel
Pungent
Phenol
Tropical
Tree fruit
Fruity
Maillard
Terpene

100 %

A
B
C
D
E
F
G
H
J
K
L
M
N
O
P
R
S
T
V

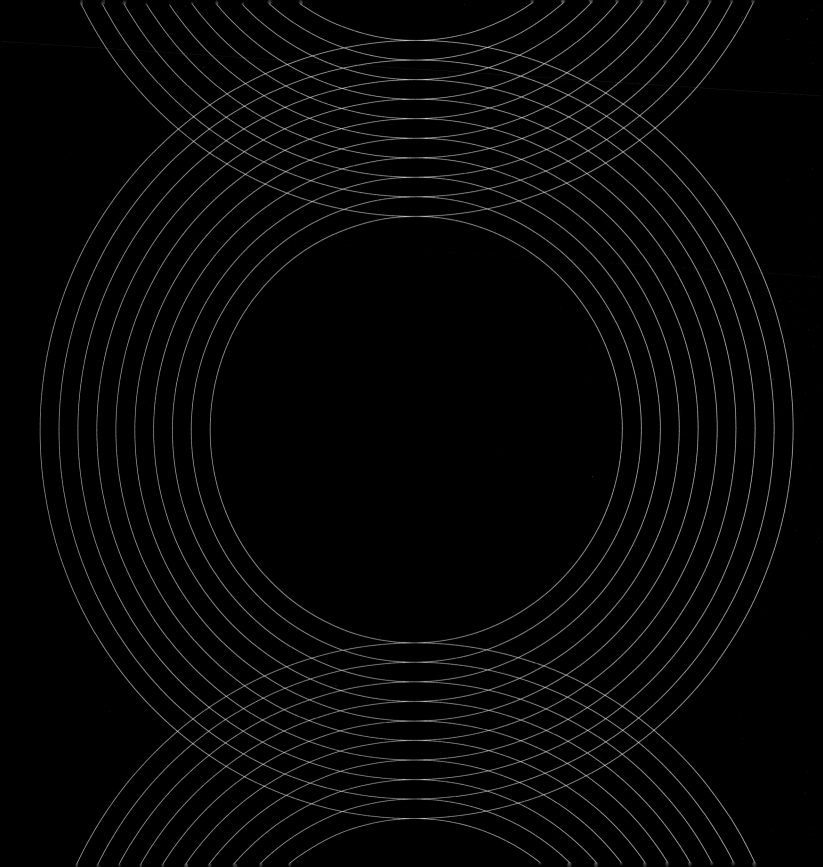

# Black Tea–Tomato Sauce

The flavors of tea and tomato are an excellent match. Black tea in particular has a deep savory edge, with herbaceous and phenolic aromas—all of which make it the ideal match for tomatoes. Tea is an extremely powerful ingredient, however, so be sure to use it judiciously—just a pinch can lend the perfect seasoning to a pan of tomato sauce. Use this as you would any tomato sauce.

**2 tablespoons olive oil**
**1 cup minced yellow onion (1 small onion)**
**4 cloves garlic, minced**
**Kosher salt and freshly ground black pepper**
**One 28-ounce can diced or crushed tomatoes**
**Loose black tea**
**1 cup water**
**Pinch of sugar (optional)**

Heat the olive oil in a medium saucepot. Add the onions and garlic and season with salt and a few grinds of pepper. Cook over medium heat until tender but not browned, about 5 minutes.

Add the tomatoes, ½ teaspoon tea, and the water. Bring to a boil and season again to taste with salt and pepper. Reduce the heat to a simmer, cover the pot, and cook for 15 minutes, until thickened.

Taste the sauce a final time and adjust the seasoning with salt and another pinch of tea and a little sugar if desired. You can purée the sauce in the pot with an immersion blender, or use it as is. Use immediately. Or transfer to a glass jar or plastic container, let cool, seal tightly, and store in the refrigerator for up to 2 weeks, or in the freezer for up to 6 months.

**MAKES 4 CUPS**

**Main Subtypes:**

Mango, pineapple, banana, coconut, passion fruit, papaya

**Best Pairings:**

Peach, vanilla, thyme, honey, cream

**Surprise Pairings:**

Lamb, dill, blue cheese, mustard

**Substitutes:**

Any tropical fruit may be substituted for another

Many of the aromatic fruits that grow in tropical regions throughout the world have very similar flavor profiles, allowing us to group them together for the purposes of matching them with other ingredients. Tropical fruit flavors are mainly created by lactones, to which many sunscreens and lotions also owe their distinctive aromas. One specific lactone, methyl hexanoate, is a common thread among all tropical fruit, and also binds them to other ingredients; it plays a significant role in the flavor of passion fruit, papaya, kiwi, oysters, pineapple, white wine, cider, vacciniums, melon, olives, raspberries, and strawberries. It also contributes to the aroma of blue cheese—and thus is one of the main factors that contribute to the surprising pairing of pineapple and blue cheese.

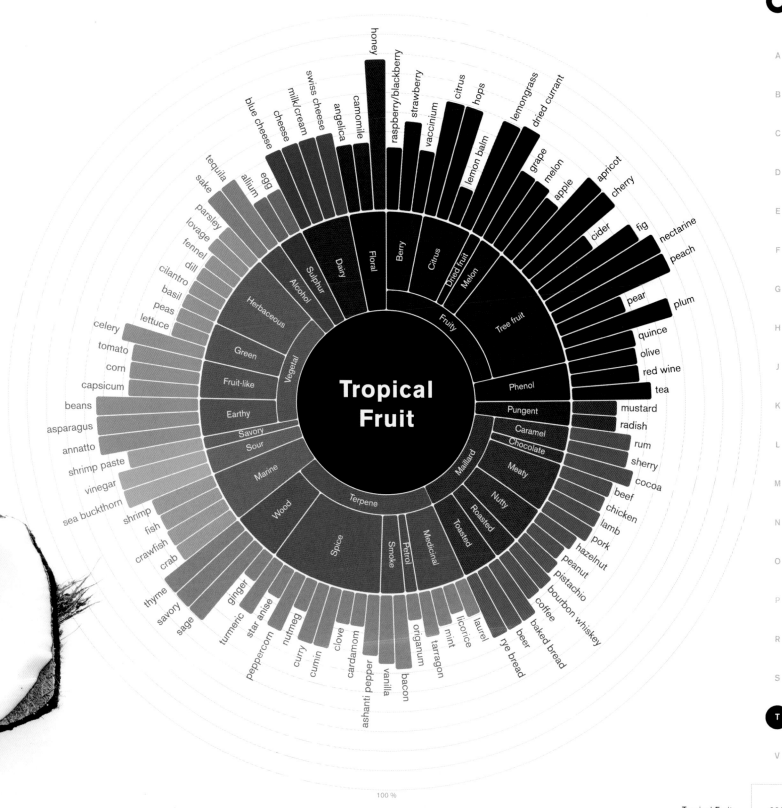

Tropical Fruit

# Crab, Mango, Dill, and Poblano Salad

Mango and dill make for a surprising and delicious combination, thanks mainly to the woody, piney aromas of both ingredients. Combined with crab, poblano, and cucumber, they create a remarkable dish that hits all of the right notes.

Grated zest and juice of ½ lemon

2 teaspoons chopped fresh dill

2 tablespoons mayonnaise

1 poblano chile, roasted, peeled, seeded, and chopped

Kosher salt and freshly ground black pepper

1 pound lump crabmeat, picked over for shells
and cartilage

1 cup diced seeded cucumber

1 mango

¼ cup celery leaves

Combine the lemon zest and juice, dill, mayonnaise, and poblano in a large bowl. Stir to combine. Season to taste with salt and pepper. Gently fold in the crab and cucumber. Cover and refrigerate until you're ready to serve, up to 1 day in advance.

Divide the salad among four plates. Peel the mango and cut the flesh from the pit. Slice the mango flesh into long pieces, ¹⁄₁₆ to ⅛ inch thick. Curl each slice and place on top of the salad. Garnish with celery leaves before serving.

**SERVES 4**

**Best Pairings:**

Grain, rice, mushroom, asparagus, egg, tomato, butter, cheese, toasted nuts

**Surprise Pairings:**

Beet, strawberry, vaccinium

**Substitutes:**

There is no direct replacement for truffle aroma; substitute mushrooms such as porcini and hen of the woods

Truffles are the most highly prized fungus in the culinary world. They grow just below the surface of the ground near the roots of certain trees, with the most desirable species—*Tuber melanosporum* (black truffles) and the rarer *Tuber magnatum* (white truffle)—growing naturally throughout southern France, Italy, and Croatia. Truffles only grow under certain conditions at specific times of the year and must be located by specially trained dogs or pigs. Their scarce supply and high international demand helps explain why truffles can fetch hundreds or even thousands of dollars per pound. The complex flavor of fresh truffles is largely created by sulfur compounds and sweet fruit aromas. White truffles should not be cooked; they lose their aroma quickly after slicing, and are best shaved directly onto food immediately before eating. Fresh black truffles release more of their aroma when heated and should be added to dishes in the last moments of cooking. If you're purchasing truffle products such as canned truffles, truffle butter, or truffle oil, always check the labels to ensure the ingredient list contains *Tuber melanosporum* or *Tuber magnatum* rather than lesser species or imitation flavors.

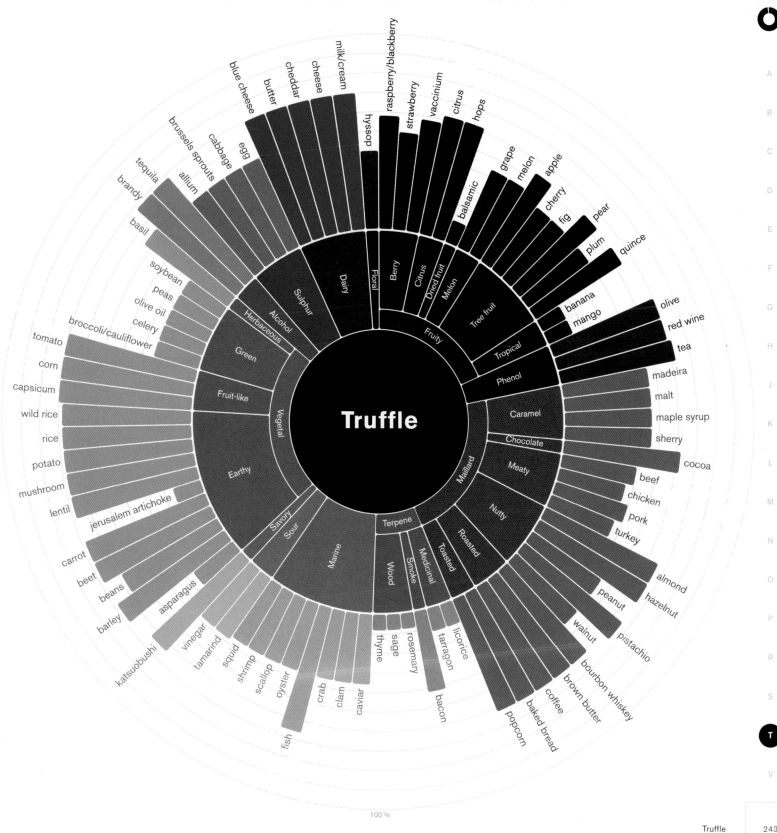

Truffle

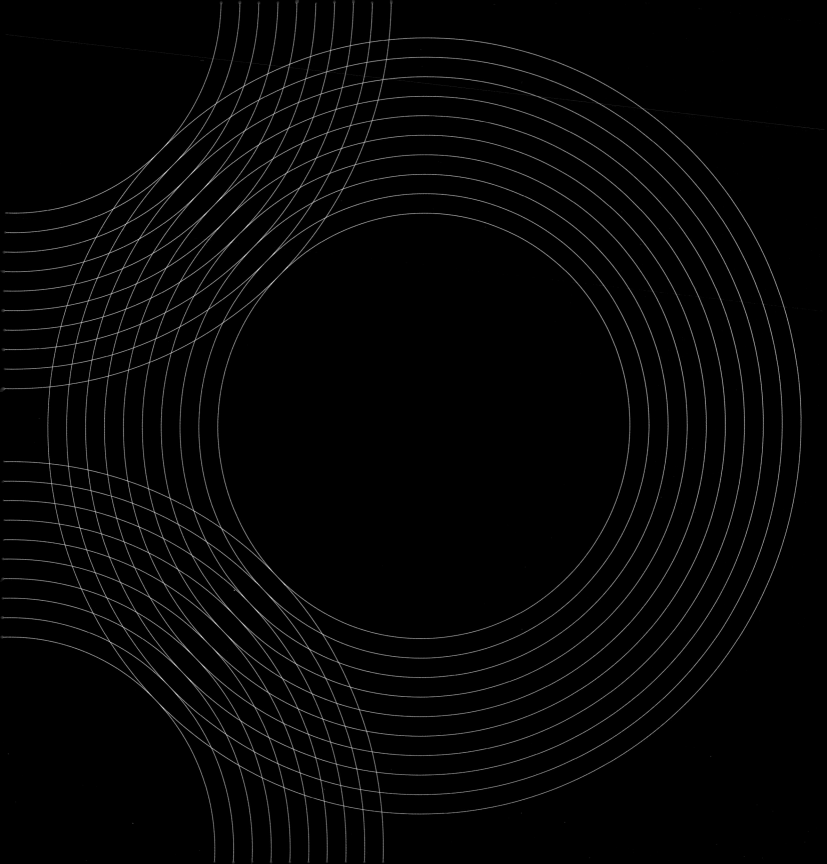

# Truffle and Roasted Beet Salad

The flavor of truffles is a combination of sulfur and fruit aromas. In this recipe, they blend perfectly with the earthy and fruity aromas of beets. Blue cheese accentuates the truffles' pungency, but may be replaced with a milder goat cheese or Robiola.

2 pounds medium-sized red beets
2 tablespoons olive oil
Kosher salt and freshly ground black pepper
4 branches fresh thyme
2 bay leaves
½ cup red wine vinegar
3 tablespoons black truffle peelings

## GARNISH

Crumbled blue cheese
Toasted walnuts or pistachios
Pea shoots or tendrils

Preheat the oven to 350°F.

Trim and scrub the beets; toss with the olive oil and salt and pepper. Place the beets in a deep baking dish. Add the thyme, bay leaves, vinegar, and 2 tablespoons of the truffle peelings. Cover with aluminum foil and roast in the oven until tender, 45 to 60 minutes.

Remove from the oven, uncover, and set the beets aside to cool slightly. When cool enough to handle, but still hot, peel them and cut into wedges or slices. Transfer to a large bowl. Strain the juices in the baking dish over the beets. Add the remaining truffle peelings and toss well to coat. Season to taste with salt and pepper and toss again. Divide the salad among four plates and garnish with blue cheese, nuts, and pea shoots before serving.

**SERVES 4**

Vanilla is a flavoring derived from the fruit of a specific genus of orchid native to Mexico. Vanilla pods (commonly called vanilla beans) must be hand harvested, because different pods on the same plant will ripen at different times. When vanilla pods are ready to harvest, they look like plump green beans. After harvest, the vanilla pod must go through curing to develop its flavor—a process that involves sweating, drying, and aging for up to six months. The cured vanilla pods are then graded using systems that vary from country to country, but that are typically based on the length, appearance, and moisture content of the pods. The aromatic compound 4-hydroxy-3-methoxybenzaldehyde, also known as vanillin, is mostly responsible for the characteristic aroma of vanilla. Vanillin is used to create artificial vanilla flavoring and "vanilla-scented" products. However, real vanilla aroma is the result of a combination of over 250 compounds.

**Best Pairings:**

Milk, cream, peach, nectarine, citrus, lemongrass, whiskey, mint

**Surprise Pairings:**

Tomato, corn, asparagus, fish

**Substitutes:**

Tonka beans may be substituted for fresh vanilla beans; substitute almond extract or bourbon whiskey for vanilla extract

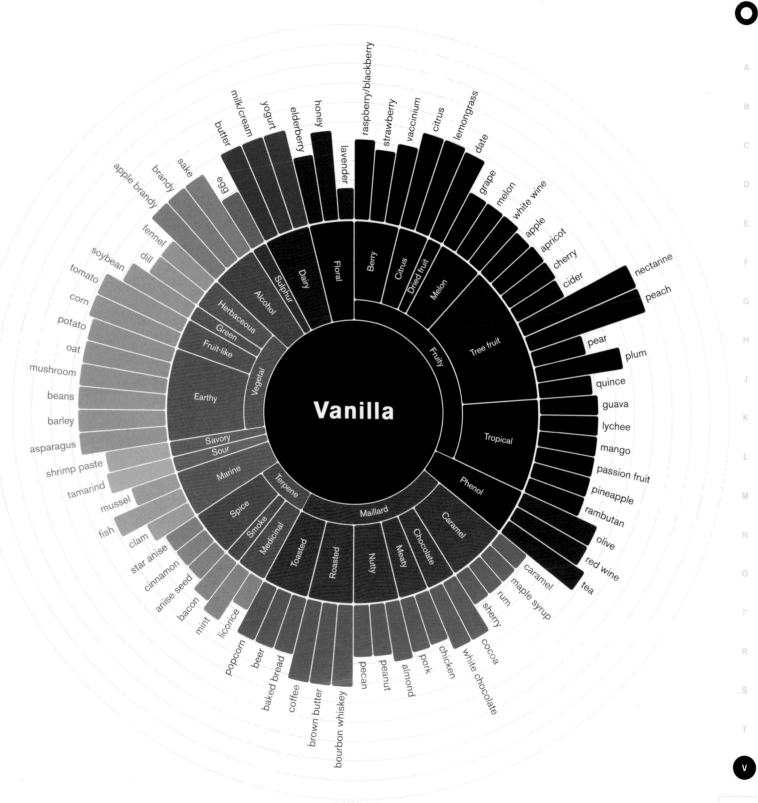

100 %

# Vanilla Butter

Vanilla is too often relegated to just the final course of a meal—which is a shame, because its incredible aroma makes it an enticing flavoring for many foods. We like to use this savory vanilla-scented butter to finish vegetables, roasted meat, and seafood—for example, we brush it over shrimp as they come off the grill, or slather it over an ear of grilled corn. (There's a reason we're focusing on the grill here: Vanilla really deepens the smoky, roasted flavors that are intrinsic to this method of cooking.)

½ pound (2 sticks) unsalted butter, softened
Seeds scraped from ½ vanilla bean (save the pod for
    another use)
2 teaspoons kosher salt
½ teaspoon freshly ground black pepper
½ teaspoon freshly ground coriander

Combine all the ingredients in a small bowl and mix until smooth and well blended. Transfer to an airtight container and store in the refrigerator for up to 4 weeks, or in the freezer indefinitely.

**MAKES 1 CUP**

# INSPIRATION

# The Elements of Flavor

# The Six Basic Tastes

Although we often use the words "taste" and "flavor" interchangeably, they actually mean something importantly different. *Taste* refers to the perceptions we experience thanks to chemical reactions that take place on the tongue when food enters the mouth. *Flavor* is the result of aromas found in food, and can be incredibly complex. Taste is only responsible for about 20 percent of what we perceive in a bite of food, but is still a very important part of the equation. Here, we list the basic tastes, and describe a few "honorary" ones, as well.

## Bitter

Bitterness is the most sensitive of all tastes because it's the most likely to save your life. Nearly all toxic substances have a pronounced bitter taste, so your sensitivity is a safeguard against ingesting poison. With that said, some bitterness is desirable, as in coffee, tea, plants in the Brassicaceae family, olives, some nuts, and cocoa.

## Sour

Sourness can be an indication of spoilage or underripeness — and thus a sign that a food should be avoided. But like bitterness, certain sour tastes in food can also be desirable, especially in dishes where a bit of tartness can counteract richer tastes like sweet or fat.

## Sweet

Our bodies like sweetness. When we detect it, it means that a food contains simple carbohydrates, sugars that can quickly be converted to fuel for our brain and muscle activity. The sensation of sweetness also sets off a series of chemical reactions that releases dopamine in the brain, which is why sweet treats literally make us happy.

## Salt

Salt receptors play a role that reaches well beyond the mouth. Without salt, cells can't function properly. Yet too much or too little salt can have negative effects on our bodies, ranging from mild muscle cramps to kidney failure. Thus, as your body senses salt intake, it sets off signals to regulate fluids accordingly. Salt also "wakes up" the taste buds in your mouth, enhancing your perception of other tastes.

## Umami

Our bodies' fifth taste receptor, for the taste known as umami, was discovered around 100 years ago by a Japanese chemist, and has gained widespread recognition in America over the past eight years. Umami (commonly translated as "pleasant savory taste") is detected by dedicated glutamate receptors on the tongue, and while it is generally associated with monosodium glutamate, or MSG, which comes from the naturally occurring amino acid glutamic acid. MSG may be added to foods to enhance flavor, and it is naturally present in foods like mushrooms, tomatoes, Parmesan cheese, soy sauce, and meat.

## Fat

The idea of fat as a taste, rather than simply a texture, remains somewhat controversial. Fat is easily recognized in the mouth because it coats the tongue and provides a feeling of unctuousness. If this were the only way we perceived fat, counting it as a taste would qualify all textures as unique tastes, blurring the definition of taste to the point of meaninglessness. However, recent research has found that the tongue can perceive free fatty acids, the compounds that make up dietary fats. There has been a documented receptor cell response to fatty acids—and historically, receptor cell detection in the mouth has been the criteria for defining tastes. So while research is still ongoing, a growing number of signs point to fat as the sixth taste.

# Other "Tastes"

The following "tastes" are commonly recognized as such because they are sensed in the mouth. But these tastes do not have dedicated chemical receptors; instead, they are sensed by the ordinary nerves in the mucous membranes inside our mouth. They are tastes without corresponding taste buds.

### Spicy

Capsaicin, an odorless (and therefore flavorless) chemical compound found in all chiles, creates the sensation of spiciness—a burning sensation—when it comes in contact with any mucous membranes. Though the burning sensation is first perceived on the lips and tongue, it is not a function of taste receptors; rather, it is carried to the brain by the same nerves responsible for movement and sensation in the face. The only way to relieve the burning is to wash the capsaicin off the tongue—and because capsaicin is fat soluble, whole milk always provides much more relief than ice water.

### Astringency

Sometimes perceived as sourness or tartness, astringency creates a dry feeling in the mouth. The dry sensation is created in part by tannic acids, which bind to salivary proteins called mucins in the mouth, disrupting their natural lubricating properties and resulting in a rough, dry sensation on the tongue. Astringency is often associated with tea and wine, but drinks with higher alcohol content can be astringent as well. Tannic acids are also found in cranberries, pomegranate, walnuts, and underripe fruit.

potatoes, a silky butternut squash purée, even yogurt or mayonnaise lend creaminess to a dish. Creamy textures provide richness and a pleasant mouthfeel, while the viscosity of creamy-textured foods also can allow the taste and flavor of foods to linger on the palate.

## Chewiness (Meatiness)

Chewiness is often thought of as an undesirable quality, yet the chewiness of meat, grains, and other vegetables are essential to plate composition. The act of chewing also helps to release the aromatic compounds that are so essential to creating flavor.

## Crispness

Crispness or crunchiness is the off-speed pitch of the texture world. It changes things up and gets the mouth and brain re-engaged with what they're perceiving. Crispness is a sign that something has been cooked well and the brain associates it with desirability, as demonstrated by experiments in the Crossmodal Research Laboratory at the University of Oxford.

## Freshness

Freshness is really a specific form of crispness. It can be the light, tender crunch of a leaf of arugula, or the hearty snap of a shaved radish. Fresh textures come from fresh vegetables, which signal wholesomeness to the brain (also demonstrated by the Crossmodal Lab) while also adding appealing colors to the plate. Fresh raw vegetables also contain aromatic compounds that are destroyed by heat, allowing us to perceive flavors that we cannot get from cooked food.

There are thousands of aromatic compounds, but in my research—and in this book—I focused on those that are most commonly found in food. The aromatic compounds found in food can be grouped based on their shapes and structures, which dictate the kinds of odors that characterize them. Each of the groups that I describe here highlights one specific chemical building block. Some chemicals are simple enough that they are only made from one of these building blocks. However, most of the chemicals I highlight consist of multiple chemical building blocks.

## Aromatic Hydrocarbons

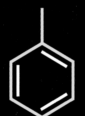

**METHYLBENZENE** (TOLUENE)
FOUND IN BEER, ANNATTO SEED, GINGER

The aromatic hydrocarbons that we find in our food tend to have six carbon atoms joined together in a circular formation called a phenyl group. Aromatic hydrocarbons tend to have solvent (paint thinner) or sweet aromas, like benzene (solvent), toluene (paint, solvent), styrene (balsamic, rubber). Their distinct aroma, which originally distinguished them from other compounds, is an essential component of flavor in many foods. Aromatic hydrocarbons may be found in everything from garlic to whiskey.

## Alcohols

**BENZYL ALCOHOL**
FOUND IN APRICOTS, CLOVE, CHICKEN, MUSTARD

An alcohol is a compound with a hydroxyl (OH) group attached. An alcohol can be something simple (such as ethanol, which contains only two carbon atoms) or more complex (such as benzyl alcohol, which contains a phenyl and an OH building block). The alcohol group contains some well-known scents; examples of specific compounds include menthol (peppermint), linalool (floral, spicy), and ethanol (rubbing alcohol).

## Carbonyls

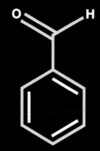

**BENZALDEHYDE** (ALDEHYDE)
IMITATION ALMOND FLAVOR

## Phenols

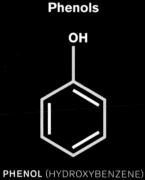

**PHENOL** (HYDROXYBENZENE)
FOUND IN CINNAMON, COFFEE, SAKE

Phenol is different from benzyl alcohol in that the OH group is directly attached to the phenyl group without a carbon atom in between. It may seem like a small distinction, but we can smell the difference between the two. Many phenols have an astringent, bitter taste, while their aroma can be sharp and pungent. You might know phenols as the bitterness in raw walnuts, cranberries, and kale; or as the tannins found in wine and tea. Other phenolic compounds range from eugenol (the main odorant in cloves) and thymol (thyme) to raspberry ketone (raspberries) and guaiacol (the smoky flavor found in coffee, whiskey, and vanilla).

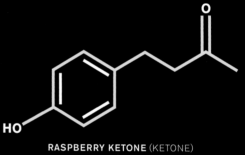

**RASPBERRY KETONE** (KETONE)
RASPBERRY FLAVOR

Carbonyls are compounds that contain a carbon-oxygen double bond. These compounds are divided into important aromatic groups called aldehydes and ketones. Aldehydes and ketones are part of some the most recognizable scents in food: vanillin (vanilla), benzaldehyde (imitation almond flavoring), carvone (spearmint extract), and cinnamaldehyde (cinnamon extract).

## Acids

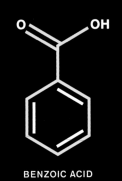

**BENZOIC ACID**
FOUND IN HONEY, NUTMEG, THYME

An acid building block is characterized by a carbonyl that is directly attached to an OH group. Their aromas vary from sour to grassy to buttery. Most aromatic acids are derived from phenols, aldehydes, or amino acids. Acids play a role in a variety of scents, such as butyric acid (vomit and rancid butter), acetic acid (vinegar), and cinnamic acid (cinnamon).

## Esters

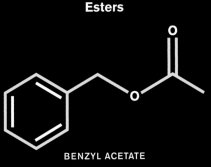

**BENZYL ACETATE**
FOUND IN PEAR, BLUEBERRY, JASMINE

Esters are similar to acids, but the hydrogen of the OH group is replaced with a different chemical building block. They are naturally found in the essential oils of plants. The vast majority of esters have strong fruity aromas, though some smell of pine or spearmint. Isoamyl acetate (banana), methyl butyrate (apple), octyl acetate (orange) and allyl hexanoate (pineapple) are common esters.

## Sulfur Compounds

**DIMETHYL SULFIDE**
FOUND IN EGG, ASPARAGUS, SOUTHERN PEAS

For most, sulfur immediately brings to mind the not-so-appealing scent of boiled eggs. However, sulfur compounds such as dimethyl sulfide (cabbage) and allyl methyl sulfide (garlic) play an important role in food items from alliums (onions, garlic, and chives) and brassicas (cabbage, cauliflower, broccoli, and Brussels sprouts) to citrus fruit and cheeses. Sulfur compounds are also part of the "roasted" aroma in chocolate, coffee, baked bread, popcorn, and roasted meat.

## Amines

**CADAVERINE**
FOUND IN BLUE CHEESE, FISH, SAUERKRAUT

Derived from ammonia, amines tend to have strong, often unpleasant scents. Putrescine (named for its putrid smell) and cadaverine (the scent of rotting flesh) are both amines. As foul-sounding as these compounds may seem, they are also found in fresh fish, beer, mushrooms, cheese, and wine. Remember, since each compound is just a pixel in a larger picture, these compounds are not individually distinguishable in any of these products, but they contribute to the overall pleasant scent of each. For example, the amine called indole has a fecal and mothball aroma, but it is an essential component in the smell of jasmine and contributes to the pairing of pork liver and jasmine.

## Pyridine and Pyrazines

**PYRIDINE**

FOUND IN ROASTED MEAT, POTATO CHIPS, BONITO FLAKES

Pyridines and pyrazines are characterized by phenyl groups in which some of the carbon atoms in the ring have been replaced with nitrogen atoms. The aroma of these compounds can range from earthy and fresh-cut grass to cocoa. Pyridines and pyrazines are the by-products of Maillard reactions (see page 261). Compounds like 2-acetylpyridine (popcorn, toasted nut) and 2-vinylpyrazine (toasted nuts) contribute to the distinctive smell of roasted meat, toasted nuts, and coffee.

## Lactones

**GAMMA-DODECALACTONE**

FOUND IN STRAWBERRIES, MUSHROOMS, CHICKEN, PORK

Lactones are esters in which the oxygen is bonded into a ring structure. They are used extensively in the flavor and fragrance industry. Most lactones have intense fruity aromas like coconut, peach, and apricot. Lactones are largely responsible for the "tropical" scent of your favorite suntan lotion. In food, they are found in tropical fruit, strawberries, mushrooms, chicken, pork, and milk products. Lactones found in food include delta-decalactone (apricot, coconut), (Z)-dairy lactone (peach, sweet) and gamma-dodecalactone (apricot, peach, flowers).

## Furans

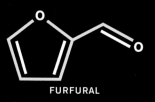

**FURFURAL**

FOUND IN BEEF, COFFEE, RUM, TAMARIND, VANILLA

Furans are molecules with a ring consisting of four carbon atoms and one oxygen atom. They tend to have a nutty aroma, and are a by-product of Maillard reactions. Furfural is one of the most common furans found in food, with the aroma of baked potato, almond, and baked bread; 2-ethylfuran has the aroma of butter and caramel, and is found in cooked meat and nuts.

The flavor matrixes in this book break down ingredients across the spectrum of aromas and flavors commonly found in foods; the divisions found in the matrixes and detailed here are of my own creation. You may find similar breakdowns of aromas in tasting wheels for coffee, beer, and wine. The categories I created were carefully selected to help you think about ingredients in a new and different way. Remember that these aromas are where the flavor of a given ingredient comes from. Therefore, pairings go beyond the ingredients listed on each matrix for a given aroma; they extend to all ingredients that share that particular aroma. Read on to better understand these aroma categories.

### Floral

Floral aromas and fruit scents are very closely related because all fruit-bearing plants produce a flower at some point in their life cycle. Floral scents tend to be more perfumy, and lack the sweet notes of fruit. These scents range from orange blossom to rose to lavender. Floral aromas are most commonly found in fruit, spices, herbs, and alcohol. For example, beta-damascenone, a key odorant in roses, can be found in apple, citrus, lemon balm, and sage; cis-ocimene, found in lilac, also plays a role in the flavor of caraway, basil, nutmeg, and thyme.

### Fruity

This is one of the broadest categories of scents. Fruity aromas include:

BERRY: This is the sweet smell associated with strawberries, blueberries, raspberries, et al. Berry aromas that come from carbonyls and esters like raspberry ketone and ethyl formate are also found in jasmine, grapes, wine, tea, tree fruit, cooked rice, and vanilla.

CITRUS: Citrus scent in its purest form is found in essential oils in the zests of lemon, lime, orange, grapefruit, and other citrus varieties. Compounds that create citrus aromas such as limonene, myrcene, and citronellol are also found in beer (via hops), olives, many herbs (especially basil, sage, mint, tarragon), ginger, fennel, and spices.

DRIED FRUIT OR JAM: The processes of drying fruit and jam making offer a double layer of aromatic compounds, first from the fruit itself (esters, lactones), which is then layered with the by-products of the drying or cooking process—such as beta-phellandrene, vanillin, and 3-ethylphenol—which create aromas of leather, smoke, and spice. Many of these compounds are also found in rosemary, cinnamon, mint, and buckwheat.

MELON: Melons are some of the most highly aromatic fruits. Scents from melon can range from sweet and fruity flesh to rinds that smell of must, musk, or rot. Among the many compounds responsible for the scent of melons, esters and aldehydes are prominent in offering aromas of fruit (melon and apple) and green (cucumber, cut grass). Similar compounds are found in cheese, citrus, rum, and tropical fruit.

TREE FRUIT: This is the group of scents associated with flowering tree fruit like apple, pear, peach, cherry, apricot, and fig. Most tree fruit are members of the Rosaceae family, meaning these fruit have similar DNA. Esters and lactones are the compounds most common in these fruit. Tree fruit aromas are also found in cocoa, olive oil, vanilla, anise, and cumin.

mas naturally found in tropical fruits like mango, pineapple, coconut, melon, and bananas. These same compounds are added to suntan lotions and mixes for blended rum drinks to create tropical aromas. These compounds can also be found in mushrooms, pork, chicken, and coffee.

## Phenol

Phenol is a specific type of aromatic hydrocarbon, but it is also a commonly used descriptor for scents. In high concentration, phenolic scents are harsh and overwhelming, smelling of tar or gasoline. In lower concentrations, their smell can be identified in aged red wines, the peat smell of Scotch, or the "tang" in pomegranate juice.

## Maillard

Louis-Camille Maillard was the French chemist who first identified the reactions that occur between amino acids and reducing sugars in browning processes. The compounds generated in these reactions are responsible for the desirable flavors in baked bread, roasted meat, and many other foods. While all aromas in this category are boosted by Maillard reactions, each has distinguishing characteristics.

CHOCOLATE: Chocolate, one of the most complex aromas, is made up of nearly 700 different aromatic compounds. The exact aroma of chocolate is determined by variable factors including the variety of the cocoa bean and the growing conditions, as well as processing (particularly what the ground beans are mixed with). However, the recognizable "chocolatey" notes are the Maillard reaction aromas associated with the roasting of the cocoa beans. The significant aromas come from aldehydes including 3-methylbutanal and vanillin as well as pyrazines. Many of these compounds can also be found in meat, as well as tamarind, tomatoes, cardamom, and eucalyptus.

MEATY, CARAMEL, TOASTED, ROASTED: Maillard reactions create hundreds of aromatic compounds including pyrroles (nutty), pyrazines (roasted), pyridines (burnt), and furans (caramel, nutty). Furfural and hydroxymethylfurfural are often pointed to as the most important compounds created in Maillard reactions. These same compounds are found in plums, coconut, popcorn, rum, and shrimp.

NUTTY: A somewhat broad subcategory of aromas, "nuttiness" includes—predictably—the smell of all nuts. Specific scents can vary depending on variety. The distinctive scent of almonds comes from a single compound. Check the label on a bottle of almond extract and you will find benzaldehyde is the only ingredient listed. Benzaldehyde, an aldehyde, is found in most nuts, and also in apricot, cherry, lemon balm, hops, corn, and sage. Some toasted aromas fit into this subcategory as well.

## Terpene

The name of this large group of strong-smelling compounds is derived from turpentine. Terpenes are described as smelling of resin, turpentine, or plastic. Found in the essential oils of a variety of plants and flowers, terpenes help create a wide range of flavors. There are over 400 naturally occurring terpenes.

MEDICINAL: Medicinal aromas are sharp, stinging scents with notes of mint. Their aromas are most associated with camphor, solvent, or mothballs. On their own, they can be somewhat unpleasant, but like all terpenes, they play a critical role in the aromas found in mint, caraway, basil, and licorice.

PETROL: "Petrol" describes another set of terpenes associated with scents that ostensibly are undesirable in food, like gasoline or kerosene, but that can play an important, positive role in the flavor of certain ingredients such as herbs like oregano, marjoram, and epazote, as well as hyssop, camomile, cilantro, and citrus.

SMOKE: These aromas are close to those of another terpene subcategory (wood) and they often originate from wood itself. As compounds in wood break down, new compounds called methoxyphenols or guaiacols are created. These compounds make the smoky aromas that result from barrel aging, smoking, charring, or roasting. They are found in significant concentration in vanilla, whiskey, Scotch, aged wines, and roasted meat and vegetables, and to a lesser extent in coffee, pork, mushrooms, corn, and honey.

SPICE: This subcategory of terpenes describes the notes created by specific spices such as black pepper, cinnamon, anise, nutmeg, and allspice, among others. There are thousands of aromatic compounds that create these aromas, and they can be found in most foods, although they are most prominent in spices, herbs, and aromatic vegetables like black pepper, turmeric, ginger, galangal, curry leaf, and cinnamon.

WOOD: These scents immediately conjure thoughts of winter holidays—fresh-cut pine, spruce, or fir. They are similar to resinous scents but have an added "fresh" or "green" aroma. They are typically found alongside pungent and/or spice scents. They are common in rosemary, sage, and ginger. They are also found in lavender, dill, fish, and berries.

## Marine

The typical aromas of seafood are best described as marine. "Fishy" has a negative connotation as an aroma descriptor, and in reality seafood should never smell fishy. Rather, it should simply smell of the sea. Sulfur compounds and amines are most common in this group. They are found in lean and fatty fish, mollusks, and crustaceans.

## Sour

Sour-smelling compounds can "sting" the nose in high concentrations, as when taking a big whiff of vinegar. Sour aromas come from acids and are common in dairy products, especially as they begin to age—in both good and bad ways. Soft-ripened cheeses, sour cream, and spoiled milk are all associated with sour. They are also found in cocoa, coffee, alcoholic beverages, and bread.

## Savory

The one category of flavor that does not rely on aromatic compounds because glutamate (the basis of umami) is odorless and sensed only on the tongue. However, savory flavors are essential to the enjoyment of food and have been shown to be effective salt replacers. These flavors are associated with cooked tomatoes, roasted mushrooms, Parmesan cheese, seaweed, fish sauce, and soy sauce.

## Vegetal

Aromas found in vegetables break down into four main subcategories:

EARTHY: Earthy aromas are often described as smelling of dirt, the forest floor, mushrooms, or mold. On the more complex end of the spectrum, these aromas also have notes of tobacco or leather. When it comes to ingredients, such smells—and the aromatic compounds that create them—are typically found in mushrooms, beets, truffles, asparagus, and malt. Among the compounds that create earthy aromas are pyrazines and pyridines, the same compounds that form the basis for the aromas of caramel, roasted meat, and nuts.

FRUIT-LIKE: Fruit-like aromas come mainly from vegetables that are technically classified as fruit. Many have a slightly sweet aroma, or simply lack the earthy and green aromas of most vegetables. The primary vegetables in this subcategory are tomato, capsicums, pumpkin, avocado, eggplant, and corn. Major compounds found in this category include 3-methylbutanal (tomato), 2-isobutyl-3-methoxypyrazine (capsicum) and hexanol (eggplant).

GREEN: Green is a common descriptor for wines made from underripe fruit. In food, green refers the crisp, somewhat tart aroma of green apple, bell pepper, or grass. As in wine, the scent is common in underripe fruit and green leaves, although the compounds responsible for green (or grassy) aromas, (Z)-3-hexenal and (E)-2-hexenal, are among the most commonly found compounds in all foods. They can be found in most herbs and green vegetables, as well as in tea, tomatoes, bananas, and cardamom.

HERBACEOUS: This vegetal aroma occupies the blurry scent space between green and floral. Woody herbs like rosemary, thyme, and sage belong in a different category of aromas, terpene; in the vegetal/herbaceous subcategory, by contrast, we find "softer" herbs like parsley, dill, chive, and clover. Herbaceous aromas can also be found in hay and in cereals like oat, bran, and dried corn. Herbaceous compounds such as thymol (thyme), estragole (fennel), and geranyl acetate (chervil) are also found in tropical fruit, mustard, chicken, and cumin.

## Alcohol

Alcoholic beverages vary widely in their aromas and flavors depending on the quantity of alcohol (ethanol) they contain, how they are distilled, and how they are aged. As a result of these differing characteristics—like smoky mezcal, roasted bourbon, or fruity white wine—you will find alcoholic beverages listed under the alcohol category as well as in other places on the matrix. The matrix also shows where ethanol is a dominant aroma, such as with vodka, brandy, and tequila, among others.

## Sulfur

The smell of sulfur is typically associated with hard-boiled eggs. The strong smell of eggs is often a result of overcooking; as the egg cooks, amino acids break down and release sulfur. This smelly reaction is also responsible for the greenish ring around the egg yolk. While the strong smell of sulfur is not desirable on its own, compounds like dimethyl disulfide, hydrogen sulfide, and benzothiazole play an important role in the flavor of brassicas, alliums, corn, cheeses, and truffles.

## Dairy

All dairy products have a similar aromatic profile at their base. After all, yogurt, crème fraîche, cheese, and even butter begin as milk or cream. Aromatic acids play a major role in the flavor of dairy products, giving even the freshest dairy products a pleasant, slightly sour smell. The distinctive aromas of dairy are also prominent in corn, pistachios, coconut, and whiskey.

## Pungent

This descriptor is often used to describe any strong, sharp scent, but it is used more accurately to describe the aromas associated with spiciness or heat in food. Mustard, horseradish, and wasabi are typical pungent ingredients. (Chiles do not fit in this category, as their aromas tend to be more fruity and vegetal. Capsaicin, the compound responsible for the heat in chiles, is odorless—and hence flavorless.) Pungent aromas are partly based on sulfur compounds and can be found in turnips and rutabagas, Brussels sprouts, arugula, watercress, radishes, and broccoli.

# Tastes, Primary Aromas, and Flavor-Pairing Charts for Each Ingredient

In case you are hungry for more details about the ingredients covered in Part II, in this section I list their tastes and key aromatic compounds. I also dive deeper into the "surprise pairings" listed in that earlier part of the book, identifying the key compounds and main aromas that characterize those pairings, as well as the tertiary ingredients that share the same key compounds—and that can therefore be added to these flavor combinations to make them even more well rounded and innovative.

## Allium

**Taste:** bitter, astringent

**Main aromas:**

Dimethyl sulfoxide—garlic

dipropyl disulfide—cooked meat, garlic, onion, pungent, sulfur

methyl propenyl disulfide—leek

pentanal—almond, bitter, malt, oil, pungent

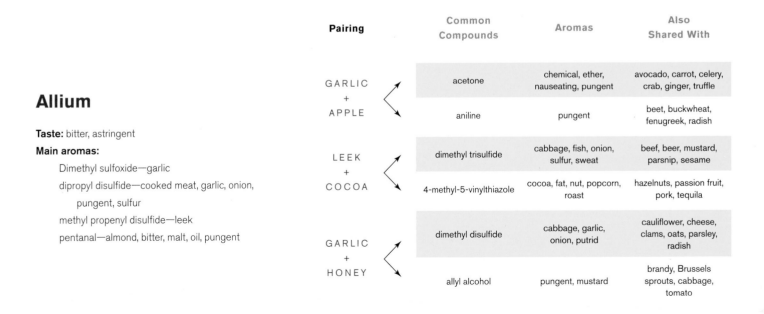

| Pairing | Common Compounds | Aromas | Also Shared With |
|---|---|---|---|
| GARLIC + APPLE | acetone | chemical, ether, nauseating, pungent | avocado, carrot, celery, crab, ginger, truffle |
| | aniline | pungent | beet, buckwheat, fenugreek, radish |
| LEEK + COCOA | dimethyl trisulfide | cabbage, fish, onion, sulfur, sweat | beef, beer, mustard, parsnip, sesame |
| | 4-methyl-5-vinylthiazole | cocoa, fat, nut, popcorn, roast | hazelnuts, passion fruit, pork, tequila |
| GARLIC + HONEY | dimethyl disulfide | cabbage, garlic, onion, putrid | cauliflower, cheese, clams, oats, parsley, radish |
| | allyl alcohol | pungent, mustard | brandy, Brussels sprouts, cabbage, tomato |

| Common Compounds | Aromas | Also Shared With | Pairing |
|---|---|---|---|
| eugenol | burnt, clove, smoke, spice | pork, savory, tarragon, anise, ginger | ARTICHOKE + PLUM |
| benzyl alcohol | rose, sweet, moss, baked bread | chile, dates, mint, radish, walnuts | |
| furfural | almond, baked potatoes, bread, burnt, spice | beef, cherries, mustard, potato | ARTICHOKE + YOGURT |
| diacetyl | butter, pastry, rancid, yeast | olives, peanuts, peas, tomato, vinegar | |
| 2-acetylthiazole | nut, popcorn, roast, sulfur | beer, clams, corn, shrimp | ARTICHOKE + SESAME |
| heptanal | citrus, dry fish, fat, green, nut, rancid | alliums, cilantro, fish, lamb | |

## Artichoke

**Taste:** bitter, lightly sweet

**Main aromas:**

1-octen-3-one—earth, metal, mushroom

1-hexen-3-one—cooked vegetable, green, metal

decanal—floral, fried, orange peel, penetrating, tallow

| Common Compounds | Aromas | Also Shared With | Pairing |
|---|---|---|---|
| 1-octen-3-ol | cucumber, earth, fat, floral, mushroom | apple, cheese, chicken, basil, truffle | ASPARAGUS + BEER |
| 4-vinylguaiacol | clove, curry, smoke, spice | citrus, coffee, fish, lovage, rice | |
| methylpyrazine | cocoa, green, hazelnut, popcorn, roasted, toasted nut | barley, beans, cabbage, shrimp, bread | ASPARAGUS + COCONUT |
| phenylacetaldehyde | berry, geranium, honey, nut, pungent | apricot, elderberries, melon, parsley, tamarind | |
| 2,6-dimethylpyrazine | cocoa, coffee, green, roast beef, toasted nut | cheese, clams, nuts, pork | ASPARAGUS + POTATO CHIPS |
| 2-acetylpyrrole | bread, cocoa, hazelnut, licorice, walnut | coconut, mushrooms, soy | |
| 2-phenylethanol | fruit, honey, lilac, rose, wine | chicken, cocoa, lettuce, olives, tomato | ASPARAGUS + MUSTARD |
| 1-pentanol | balsamic, fruit, green, pungent, yeast | apple, lentils, mushrooms, sherry | |

## Asparagus

**Taste:** sweet, slightly bitter

**Main aromas:**

dimethyl sulfide—cabbage, gasoline, sulfur, wet earth

hexanol—cooked vegetable, flower, green, herb, rot

3-hydroxy-2-butanone—butter, cream, green bell pepper, rancid, sweat

## Avocado

**Taste:** fat

**Main aromas:**

hexanal—fresh, fruit, grass, green, oil

(E)-2-hexenal—fat, floral, fruit, grass, pungent

2,4-hexadienal—cut grass, earth, fish, green

| Pairing | Common Compounds | Aromas | Also Shared With |
|---|---|---|---|
| AVOCADO + TEQUILA | 2-methyl-1-propanol | apple, bitter, cocoa, plastic, solvent | alliums, brassicas, corn, mango |
| | methyl heptenone | citrus, mushroom, black pepper, rubber, strawberry | annato, ginger, lemon balm, tomato |
| AVOCADO + COCOA | acetoin | butter, cream, green bell pepper, rancid | lychee, passion fruit, vinegar |
| | 2-pentylfuran | butter, floral, fruit, green bean | cashews, malt, parsley, shrimp |
| AVOCADO + APPLE | 1-hexanol | banana, flower, grass, green, herb, nut, resin | lettuce, turnip, celery, dill, peas |
| | 2-ethyl-1-hexanol | citrus, green, oil, rose | cherries, dried bonito, endive, pineapple |

## Beef

**Taste:** umami, fat, sweet

**Main aromas:**

2,3,5-trimethylpyrazine—cocoa, earth, must, potato, roast

1-octen-3-ol—cucumber, earth, fat, floral, mushroom

3-methylbutanoic acid—cheese, feces, putrid fruit, rancid, sweat

heptanal—citrus, dry fish, fat, green, nuts

| Pairing | Common Compounds | Aromas | Also Shared With |
|---|---|---|---|
| BEEF + COCOA | acetophenone | almond, flower, meat, must, plastic | buckwheat, cauliflower, corn, tamarind, wine |
| | isopentanal | cocoa, cooked vegetable, fresh, green, malt, spice | beer, hazelnuts, ginger, whiskey |
| BEEF + BEER | naphthalene | mothball, tar | beans, capsicums, peanuts, sage, shrimp |
| | limonene | balsamic, citrus, fragrant, fruit, greenery, herb | apple, caraway, chervil, fig, okra, sweet potato |
| BEEF + CABBAGE | 2-methylthiophene | sulphur | alliums, citrus, milk products, popcorn, whiskey |
| | 2-pentylfuran | butter, floral, fruit, green bean | artichoke, clams, parsley, tea, tomato |
| BEEF + GRAPE | 2,4,5-trimethyloxazole | nut, sweet | chicken, cocoa, oats, potato, soy sauce |
| | pseudocumene | pesticide, plastic | coffee, crab, vanilla, walnuts, watercress |

| Common Compounds | Aromas | Also Shared With | Pairing |
|---|---|---|---|
| isoamyl alcohol | bitter, burnt, cocoa, gasoline, malt | cherries, cheese, coffee, mustard, bread | BEET + LITCHI |
| ethanol | alcohol, floral, ripe apple, sweet | alliums, carrot, crab, olives, raspberries | |
| dimethyl sulfoxide | garlic | blackberries, cucumber, mint, onion | BEET + OYSTER |
| capronaldehyde | fresh, fruit, grass, green, oil | camomile, ginger, parsley, radish | |
| phenylacetaldehyde | berry, geranium, honey, nut, pungent | apricot, elderberries, corn, sage | BEET + LEMON BALM |
| carvone | basil, bitter, caraway, fennel, mint | celery, fennel, savory, orange | |

# Beet

**Taste:** sweet

**Main aromas:**

4-methylpyridine—ash, sweat

isovaleraldehyde—almond, cocoa, cooked vegetable, malt, spice

geosmin—beet, earth

# Berry

**Taste:** sweet, sour
**Main aromas:**

STRAWBERRY

Furaneol—burnt, caramel, cotton candy, honey,
sweet; mainly responsible for toasted

mesifurane—bread crust, butter, caramel

ethyl butanoate—apple, butter, pineapple, red
fruit, strawberry

RASPBERRY

raspberry ketone—citrus, raspberry

beta-ionone—cedar, floral, raspberry, seaweed,
violet

geraniol—geranium, lemon zest, passion fruit,
peach, rose

BLACKBERRY

2-heptanol—citrus, coconut, earth, fried,
mushroom, oil

p-cymen-8-ol—citrus, must, sweet

2-heptanone—bell pepper, blue cheese, green,
nut, spice

alpha-terpinene—berry, lemon, wood

BLUEBERRY

(Z)-3-hexenal—apple, bell pepper, cut grass,
green, lettuce

(E,E)-2,4-hexadienal—citrus, fat, floral, green

cucumber aldehyde—cucumber, green, lettuce,
wax

methyl 2-methylbutanoate—apple, fruit, green
apple, strawberry, sweet

butyl acetate—apple, banana, glue, pungent,
sweet

2-methylbutyl acetate—apple, banana, pear

CRANBERRY

alpha-terpineol—anise, fresh, mint, oil, sweet

benzaldehyde—bitter almond, burnt sugar,
cherry, malt, roasted pepper

methyl 2-methylbutanoate—apple, fruit, green
apple, strawberry, sweet

| Pairing | Common Compounds | Aromas | Also Shared With |
|---|---|---|---|
| VACCINIUM + THAI BASIL | limonene | balsamic, citrus, fragrant, fruit, green, herb | celery, fennel, orange, turmeric, sesame |
| | 3-octanone | butter, mold, herb, resin | coffee, peanuts, turkey |
| STRAWBERRY + MUSHROOM | 1-nonanol | fat, floral, green, oil | chicken, cauliflower, basil, cheese |
| | Methyl butanoate | apple, banana, butter, cheese, ester, floral | apple, gooseberries, olives, wine |
| RUBUS + CUMIN | beta-caryophyllene | fried, spice, wood | carrot, star anise, pistachios, mint |
| | 1-isopropyl-4-methylbenzene (=p-cymene) | citrus, fresh, gasoline, solvent | cardamom, eucalyptus, ginger, black pepper, walnuts |

| Common Compounds | Aromas | Also Shared With | Pairing |
|---|---|---|---|
| 4-methylacetophenone | bitter almond, floral, fruit, spice, sweet | capsicums, celery, lime, passion fruit | CAULIFLOWER + COCOA |
| 2-butenal | pungent | caviar, cheese, dates, peanuts | |
| (E,E)-2,4-heptadienal | fat, fish, nut, plastic | alliums, beef, tomato, walnuts | BROCCOLI + FIG |
| indole | burnt, feces, medicinal, mothball | blue cheese, hazelnuts, okra, rice, shrimp paste | |
| (E,E)-2,4-decadienal | cilantro, deep-fried, fat, oil, oxidized | chile, clams, lime, mushrooms, yogurt | BROCCOLI/ CAULIFLOWER + PEANUT |
| dimethyl trisulfide | cabbage, fish, onion, sulfur, sweat | oyster, parsnip, pumpkin, radish | |

## Brassica Oleracea: Floral

**Taste**: slightly bitter, slightly sweet

**Main aromas:**

3-butenylisothiocyanate—green, pungent, sulfur

cis-3-Hexen-1-ol—bell pepper, grass, green leaf, herb, unripe banana

nonanal—citrus, fat, floral, green, paint

| Common Compounds | Aromas | Also Shared With | Pairing |
|---|---|---|---|
| indole | burnt, feces, medicinal, mothball | cauliflower, lemon, shrimp | KOHLRABI + PORK |
| 2,6-dimethylpyrazine | cocoa, coffee, green, roast beef, toasted nut | asparagus, barley, coffee, hazelnuts | |
| 2-butanone | ether, fruit, sweet | beer, cheddar cheese, pumpkin seeds, truffle | BRUSSELS SPROUTS + APPLE |
| methyl octanoate | fruit, orange, sweet, wax, wine | cabbage, chile, mussels | |
| 1-tetradecanol | coconut | citrus, crab, gooseberries, scallion | KALE + COCONUT |
| 2-acetylfuran | balsamic, cocoa, coffee, smoke, tobacco | beef, cocoa, oyster, pumpkin | |

## Brassica Oleracea: Leafy

**Taste:** bitter

**Main aromas:**

allyl isothiocyanate—garlic, pungent, sulfur

butyl isothiocyanate—green, pungent, sulfur

(E)-4-hydroxycinnamic acid—astringent, balsamic, phenol

## Brassica Rapa

**Taste:** bitter, mild spicy

**Main aromas:**

phenylethyl isothiocyanate—horseradish, mustard

eugenol—burnt, clove, smoke, spice

longifolene—flower, vegetable

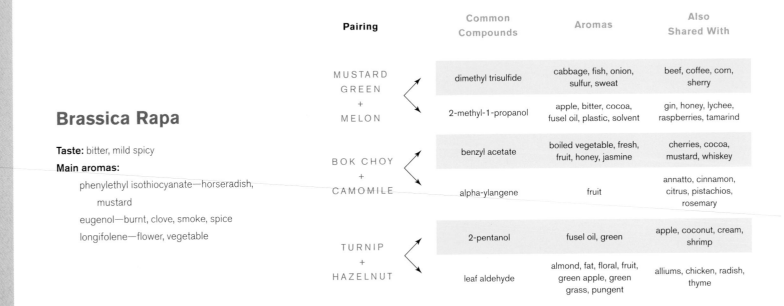

| Pairing | Common Compounds | Aromas | Also Shared With |
|---|---|---|---|
| MUSTARD GREEN + MELON | dimethyl trisulfide | cabbage, fish, onion, sulfur, sweat | beef, coffee, corn, sherry |
| | 2-methyl-1-propanol | apple, bitter, cocoa, fusel oil, plastic, solvent | gin, honey, lychee, raspberries, tamarind |
| BOK CHOY + CAMOMILE | benzyl acetate | boiled vegetable, fresh, fruit, honey, jasmine | cherries, cocoa, mustard, whiskey |
| | alpha-ylangene | fruit | annatto, cinnamon, citrus, pistachios, rosemary |
| TURNIP + HAZELNUT | 2-pentanol | fusel oil, green | apple, coconut, cream, shrimp |
| | leaf aldehyde | almond, fat, floral, fruit, green apple, green grass, pungent | alliums, chicken, radish, thyme |

## Capsicum

**Taste:** sweet, spicy

**Main aromas:**

2-isobutyl-3-methoxypyrazine—earth, floral, green bell pepper, pea

diacetyl—butter, caramel, fruit, sweet, yogurt

3-carene—bell pepper, lemon, pungent, resin, rubber

| Pairing | Common Compounds | Aromas | Also Shared With |
|---|---|---|---|
| ANCHO + PUMPKIN | 2-(sec-butyl)-3-methoxypyrazine | bell pepper, carrot, earth, green | beet, cucumber, peanuts |
| | 3-pentanone | ether, sweet | alliums, cream, mushrooms, pear, sage |
| JALAPEÑO + PEACH | pentadecane | waxy | coconut, leek, mustard, pork |
| | limonene | balsamic, citrus, fragrant, fruit, green, herb | basil, camomile, coffee, star anise |
| BELL PEPPER + DILL | camphor | camphor, earth, pine, spice | fennel, lemon balm, tomato |
| | sabinene | herb, black pepper, turpentine, wood | celery, cumin, ginger, pistachios |

| Common Compounds | Aromas | Also Shared With | Pairing |
|---|---|---|---|
| guaiacol | burnt, medicinal, phenol, smoke, wood | beer, cabbage, citrus, tomato | CARAMEL + FISH SAUCE |
| diacetyl | butter, caramel, fruit, pastry, rancid, sweet, yeast | apricot, brandy, cream, rye bread | |
| maltol | caramel, cotton candy, malt, toasted bread, toasted nut | chestnuts, cocoa, pecans, pork | CARAMEL + TAMARIND |
| 5-methylfurfural | almond, caramel, cooked, roasted garlic, spice | banana, beef, orange, potato | |
| furaneol | burnt, caramel, cotton candy, honey, sweet | mango, popcorn, soy sauce, strawberries | CARAMEL + PEANUT |
| y-decalactone | fat, fruit, lactone, peach, sweet | coconut, Parmesan, plum, sherry | |
| vanillin | sweet, vanilla | fennel, lemon, oats, pineapple | CARAMEL + APPLE |
| furaneol | burnt, caramel, cotton candy, honey, sweet | chestnuts, coffee, passion fruit, pistachios | |

## Caramel

**Taste:** sweet, slightly bitter

**Main aromas:**

5-(hydroxymethyl)furfural—caramel, cardboard, fat, must, spice

maltol—caramel, cotton candy, malt, roasted bread, toasted nut

diacetylformoin—caramel, toasted almond, butter

| Common Compounds | Aromas | Also Shared With | Pairing |
|---|---|---|---|
| myrcene | balsamic, flower, fruit, herb, must | annatto, cilantro, turmeric, rosemary | CARROT + BALSAMIC |
| 6-allyl-4-methoxy-1,3-benzodioxole | balsamic, carrot, spice, warm | black pepper, blueberries, dill, parsnip | |
| terpinolene | pine, plastic, sweet | cardamom, lime, ginger, mango | CARROT + BASIL |
| bornyl acetate | herb, pine, sweet | dried currants, cinnamon, pistachios | |
| 4-vinylguaiacol | clove, curry, smoke, spice | cider, lovage, pork, sage | CARROT + COFFEE |
| 5-methylfurfural | almond, caramel, cooked, roasted garlic, spice | coconut, corn, rum, tamarind | |

## Carrot

**Taste:** sweet

**Main aromas:**

alpha-terpinolene—pine, plastic, sweet

alpha-pinene—cedar, pine, resin, sharp, turpentine

beta-myrcene—balsamic, fruit, geranium, herb, must

## Citrus

**Taste:** sour, sweet

**Main aromas:**

CITRUS

limonene—balsamic, citrus, fragrant, fruit, herb

ORANGE

valencene—citrus, green, oil, wood

LEMON

gamma-terpinene—bitter, citrus, gasoline, resin, turpentine

LIME

beta-pinene—pine, polish, resin, turpentine, wood

GRAPEFRUIT

myrcene—balsamic, fruit, geranium, herb, must

| Pairing | Common Compounds | Aromas | Also Shared With |
|---|---|---|---|
| GRAPEFRUIT + SAGE | 1,8-cineol | camphor, cool, eucalyptus, mint, sweet | basil, camomile, cardamom, fennel, black pepper |
| | alpha-phellandrene | citrus, mint, black pepper, turpentine, wood | cinnamon, nutmeg, lovage, dill, star anise |
| ORANGE + MUSTARD | nonanal | citrus, fat, floral, green, lemon, paint, soap | annatto, beer, cilantro, hazelnuts |
| | 1-hexanol | banana, cooked vegetable, flower, grass, green, herb, lettuce, resin, rot, spice | asparagus, brassicas, cilantro, soybeans, tomato |
| LEMON + EUCALYPTUS | gamma-terpinene | bitter, citrus, gasoline, resin, turpentine | anise, caraway, ginger, pistachios |
| | tridecane | lime oil | eggplant, bay leaf, mustard seeds |
| LIME + PEACH | leaf aldehyde | almond, fat, floral, fruit, green apple, grass, pungent | cucumber, endive, olives, radish |
| | 4-hydroxydodecanoic acid, gamma-lactone | apricot, flower, fruit, peach, sweet | chervil, chicken, rum, strawberries |

## Cocoa

**Taste:** bitter

**Main aromas:**

3-methylbutanoic acid—sweat, rancid

furaneol—burnt, caramel, cotton candy, honey, sweet

2-ethyl-3,5-dimethylpyrazine—potato chips

| Pairing | Common Compounds | Aromas | Also Shared With |
|---|---|---|---|
| COCOA + BEET | phenethylamine | ammonia, fishy | blue cheese, citrus, rhubarb, sake |
| | benzaldehyde | bitter almond, burnt sugar, cherry, malt, roasted pepper | almonds, avocado, cinnamon, lemon thyme, sesame seeds |
| CHOCOLATE + PASSION FRUIT | linalool | bergamot, citrus, cilantro, lavender, lemon, rose | cilantro, fennel, mint, pistachios, tarragon |
| | 2-acetylfuran | balsamic, cocoa, coffee, smoke, tobacco | barley, jalapeño, oyster, pumpkin |
| CHOCOLATE + PEANUT BUTTER | 3-hydroxy-2-methyl-4-pyrone | caramel, cotton candy, malt, toasted bread, toasted nut | chestnuts, citrus, pineapple |
| | pseudocumene | pesticide, plastic | banana, nectarine, vanilla, walnuts |

| Common Compounds | Aromas | Also Shared With | Pairing |
|---|---|---|---|
| 2-phenylethanol | fruit, honey, lilac, rose, wine | alliums, chicken, tamarind, truffle | CORN + COCONUT |
| ethyl acetate | brandy, contact glue, grape, sweet | capsicums, fig, mussels, sake, vinegar | |
| octadecane | citrus, black pepper | citrus, lamb, mustard, pecans, rosemary | CORN + VANILLA |
| 4-hydroxybenzaldehyde | roast | apple, coffee, honey, pineapple, sherry | |
| beta-damascenone | cooked apple, floral, fruit, honey, tea | beer, elderberries, lemon balm, tea, whiskey | CORN + APPLE |
| hexyl acetate | apple, banana, confectionery, fruit, grass, herb, pear | chestnuts, oregano, mushrooms, whiskey | |

# Corn

**Taste:** sweet

**Main aromas:**

dimethyl sulfide—cabbage, gasoline, sulfur, wet earth

hydrogen sulfide—rotten egg

methanethiol—cabbage, garlic, gasoline, putrid, sulfur

| Common Compounds | Aromas | Also Shared With | Pairing |
|---|---|---|---|
| dimethyl sulfoxide | garlic | alliums, beer, beet, crème fraîche | WATERCRESS + OYSTER |
| 1-penten-3-ol | burnt, fish, green, meat, wet earth | dill, lemon, mustard, tea | |
| styrene | balsamic, gasoline, plastic, rubber, solvent | lentils, olives, peas, scallion | CRESS (ANY) + STRAWBERRY |
| pseudocumene | pesticide, plastic | bread, crab, hazelnuts, nectarine | |
| allyl isothiocyanate | garlic, pungent, sulfur | beer, cucumber, mint, blackberries | NASTURTIUM + PINEAPPLE |
| acetic acid | acid, fruit, pungent, sour, vinegar | annatto, coffee, lime, vanilla | |

# Cress

**Taste:** bitter, spicy

**Main aromas:**

phenylethyl isothiocyanate—horseradish, mustard

5-(methylthio)pentanenitrile—broccoli, cabbage

benzenemethanethiol—burnt wood, smoke

# Crustacean

**Taste:** sweet, lightly salty

**Main aromas:**

N,N-dimethyl-1-butanamine—fish

5,6-Dihydro-5-methyl-4H-1,3,5-dithiazine—fried onion, leek, meat, savory, seafood, sulfur

3-hydroxybutan-2-one—butter, cream, green bell pepper, rancid, sweat

| Pairing | Common Compounds | Aromas | Also Shared With |
|---|---|---|---|
| SHRIMP + LAMB | 4-hydroxynonanoic acid, lactone | apricot, cocoa, coconut, peach, sweet | beans, fenugreek, mushrooms, oregano |
| | 2-methylpyridine | ash, sweat | alliums, katsuobushi, oats, peanuts, rice |
| CRAWFISH + PUMPKIN | 2,3-pentanedione | bitter, butter, caramel, cream, fruit, sweet | coffee, hazelnuts, okra, rye bread |
| | 2,5-dimethylpyrazine | burnt plastic, cocoa, medicinal, roast beef, toasted nut | asparagus, barley, pork, rice |
| CRAB + CHERRY | 2-ethyl-1-hexanol | citrus, green, oil, rose | grapes, lettuce, milk, olives, tomato |
| | 1,3-dimethylbenzene | fried, medicinal, nut, plastic, rancid | avocado, dill, endive, walnuts |
| LOBSTER + BEEF | 2,4-dithiapentane | sulfur, garlic | lychee, milk, mushrooms, truffle |
| | caprylaldehyde | citrus, fat, floral, green, pungent | artichoke, olives, pomegranate, shallot |

# Cucumber

**Taste:** slightly bitter

**Main aromas:**

(E)-2-nonenal—cucumber, cut grass, fat, paper, watermelon

trans, cis-2,6-nonadienal—cucumber, green, lettuce, wax

cis-6-nonenal—melon

| Pairing | Common Compounds | Aromas | Also Shared With |
|---|---|---|---|
| CUCUMBER + BLUEBERRY | tetradecanal | flower, hay, honey, wax, wood | alliums, beer, chicken, pistachios, scallops |
| | (E)-2-nonenal | cucumber, cut grass, fat, paper, watermelon | camomile, fig, mango, pecans, tamarind |
| CUCUMBER + HAZELNUT | 2-(sec-butyl)-3-methoxypyrazine | bell pepper, carrot, earth, green | asparagus, beet, celery, parsnip, zucchini |
| | 2-pentylfuran | butter, floral, fruit, green bean | avocado, chile, pork, shrimp |
| CUCUMBER + PERSIMMON | lauric acid | fat, fruit, metal, wax | apricot, dill, lime, yogurt |
| | (Z)-2-penten-1-ol | banana, green, plastic, rubber | clams, lettuce, prickly pear, thyme |

| Common Compounds | Aromas | Also Shared With | Pairing |
|---|---|---|---|
| 2-heptanone | bell pepper, blue cheese, fruit, green, nut, spice | blue cheese, caviar, chile, hazelnuts, pear | CREAM + OYSTER |
| 1-octanol | bitter almond, burnt matches, fat, floral, metal | cauliflower, coconut, mustard, sage | |
| Hexyl methyl ketone | fat, fragrant, gasoline, mold, soap | curry, miso, truffle, vanilla, wine | YOGURT + WALNUT |
| 2-pentanone | burnt plastic, ether, fruit, kerosene, pungent | carrot, coffee, mushrooms, pumpkin seeds, shrimp | |
| 1-heptanol | chemical, green, putrid, wood | apple, chestnuts, dill, olives, sherry | MILK + PORK |
| vanillin | sweet, vanilla | asparagus, fennel, lemon, peanuts | |
| 2-undecanone | fresh, green, orange, rose | alliums, dates, lemongrass, turmeric | SOUR CREAM + CLAM |
| hexanoic acid | cheese, goat, oil, pungent, sweat | cashews, coffee, mushrooms, polenta | |

## Dairy

**Taste:** lightly sweet, sour

**Main aromas:**

2-propanone—chemical, ether, nauseating, pungent

2-butanone—ether, fragrant, fruit, pleasant, sweet

2-heptanone—bell pepper, blue cheese, green, nut, spice

acetaldehyde—ether, floral, green apple, pungent, sweet

| Common Compounds | Aromas | Also Shared With | Pairing |
|---|---|---|---|
| dimethyl sulfide | cabbage, gasoline, sulfur, wet earth | cauliflower, chile, oyster, parsley | EGG + TRUFFLE |
| 2-butanone | ether, fragrant, fruit, pleasant, sweet | chicken, potato, rice, scallops, walnuts | |
| ammonia | pungent | beet, cabbage, celery, radish, rhubarb | EGG + CAVIAR |
| 2,4-decadienal | deep-fried | dates, hazelnuts, pears, sesame seeds, tomato | |
| thiophene | garlic | crab, coffee, pork, whiskey | EGG + VANILLA |
| 1,3-dimethylbenzene | fried, medicinal, nut, plastic, rancid | avocado, basil, beans, cream, watercress | |

## Egg

**Taste:** bland; flavor is fully dependent on aroma

**Main aromas:**

hydrogen sulfide—sulfur, rotten egg

methional—cooked potato, soy

phenylacetaldehyde—berry, geranium, honey, nut, pungent

2-acetyl-1-pyrroline—nut, oil, popcorn, roast

## Eggplant

**Taste:** bitter, slightly sweet when cooked

**Main aromas:**

hexanol—cooked vegetable, flower, green, herb, rot

toluene—glue, paint, solvent

3-carene—bell pepper, lemon, pungent, resin, rubber

| Pairing | Common Compounds | Aromas | Also Shared With |
|---|---|---|---|
| EGGPLANT + WALNUT | 1,2-dimethylbenzene | geranium | artichoke, beer, lentils, southern peas, scallops |
| | acetophenone | almond, flower, meat, must, plastic | blue cheese, cocoa, corn, tea |
| EGGPLANT + POMEGRANATE | heptanal | citrus, dry fish, fat, green, nut, rancid | apricot, beef, dill, sesame seeds |
| | beta-bergamotene | tea | basil, carrot, citrus, black pepper |
| EGGPLANT + ELDERBERRY | linalool | bergamot, citrus, cilantro, floral, lavender, lemon, rose | cumin, hyssop, lemon balm, sake |
| | heptane | burnt matches, floral, plastic | chestnuts, port, thyme, yogurt |
| EGGPLANT + LAMB | linalool | bergamot, citrus, cilantro, lavender, lemon, rose | capers, celery, lemongrass, mint, peach, star anise |
| | heptanal | citrus, dried fish, fat, green, nut, rancid | cilantro, ginger, corn, pistachios, orange |

## Fennel

**Taste:** sweet

**Main aromas:**

anethole—anise, sweet

limonene—balsamic, citrus, fragrant, fruit, herb

alpha-phellandrene—citrus, mint, black pepper, turpentine, wood

| Pairing | Common Compounds | Aromas | Also Shared With |
|---|---|---|---|
| FENNEL + HYSSOP | anethole | anise, sweet | dill, cilantro, elderberries, rhubarb, shiso |
| | linalool | bergamot, cilantro, lavender, lemon, rose | camomile, celery, mint, turmeric |
| FENNEL + PLUM | 4-methoxybenzaldehyde | almond, anise, mint, sweet | anise, basil, coffee, hazelnuts, black pepper |
| | 1,8-cineol | camphor, cool, eucalyptus, mint, sweet | gin, lemongrass, passion fruit, star anise |
| FENNEL + VACCINIUM | limonene | balsamic, citrus, fragrant, fruit, green, herb | dried currants, caraway, chervil, lovage, pecans, pineapple |
| | 1-isopropyl-4-methylbenzene | citrus, fresh, gasoline, solvent | curry leaf, lime, yogurt, parsley |

| Common Compounds | Aromas | Also Shared With | Pairing |
|---|---|---|---|
| 2-phenylethanol | fruit, honey, lilac, rose, wine | cocoa, corn, mezcal, watermelon | FIG + AVOCADO |
| 3-hydroxy-2-butanone | butter, cream, green bell pepper, rancid, sweat | beef, blue cheese, cherries, olives, vinegar | |
| 6-methyl-5-hepten-2-one | citrus, mushroom, black pepper, rubber, strawberry | carrot, citrus, lemongrass, thyme, walnuts | FIG + OLIVE |
| (E,E)-2,4-decadienal | cilantro, deep-fried, fat, oil, oxidized | apricot, camomile, cauliflower, chicken, mushrooms | |
| ethanol | alcohol, floral, ripe apple, sweet | apple, grapes, peach, pineapple, pistachios, vinegar | FIG + CLAM |
| geranylacetone | green, hay, magnolia | apricot, capsicums, passion fruit, sage | |

## Fig

**Taste:** sweet

**Main aromas:**

(E)-2-hexanal—green banana, fresh, herbal, spice

5-methylfurfural—almond, caramel, cooked, roasted garlic, spice

beta-caryophyllene—fried, spice, wood

| Common Compounds | Aromas | Also Shared With | Pairing |
|---|---|---|---|
| 1-penten-3-ol | burnt, butter, fish, green, meat, oxidized, wet earth | apricot, clams, olives, tomato | LEAN FISH + MELON |
| 2,3-butanediol | fruit, herb, onion | bell pepper, dates, fig, sherry, vinegar | |
| 2-decenal | fat, fish, orange, paint | carrot, cilantro, lamb, olives, walnuts | LEAN FISH + PEANUT |
| naphthalene | mothball, tar | avocado, chile, kiwi, orange, sage | |
| phenol | medicinal, phenol, sharp, smoke, spice | cocoa, olives, pork, sake, sesame seeds | FATTY FISH + COFFEE |
| 2-ethyl-1-hexanol | citrus, green, oil, rose | annatto, lettuce, mint, pomegranate, sweet potato | |
| pyrrolidine | ammonia, fishy | barley, caviar, polenta, radish | FATTY FISH + BEER |
| formic acid | pungent | citrus, brandy, peas, yogurt | |

## Fish

**Taste:** salt, fat

**Main aromas:**

2,4,6-tribromophenol—saltwater fish, brine-like

2,6-nonadienal—cucumber, green

1,5-octadien-3-ol—earth, geranium, green, herb, mushroom

## Game Meat

**Taste:** umami

**Main aromas:**

acetone—chemical, ether, nauseating, pungent

n-butanol—alcohol, fruit, medicinal, solvent

isobutanol—apple, bitter, cocoa, fusel oil, solvent

skatole—feces, mothball

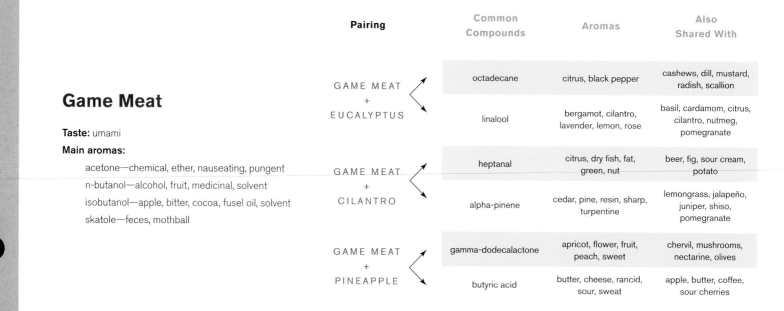

| Pairing | Common Compounds | Aromas | Also Shared With |
| --- | --- | --- | --- |
| GAME MEAT + EUCALYPTUS | octadecane | citrus, black pepper | cashews, dill, mustard, radish, scallion |
| | linalool | bergamot, cilantro, lavender, lemon, rose | basil, cardamom, citrus, cilantro, nutmeg, pomegranate |
| GAME MEAT + CILANTRO | heptanal | citrus, dry fish, fat, green, nut | beer, fig, sour cream, potato |
| | alpha-pinene | cedar, pine, resin, sharp, turpentine | lemongrass, jalapeño, juniper, shiso, pomegranate |
| GAME MEAT + PINEAPPLE | gamma-dodecalactone | apricot, flower, fruit, peach, sweet | chervil, mushrooms, nectarine, olives |
| | butyric acid | butter, cheese, rancid, sour, sweat | apple, butter, coffee, sour cherries |

## Ginger

**Taste:** sweet, spicy

**Main aromas:**

zingiberene—fresh, sharp, spice

beta-sesquiphellandrene—wood

humulene—balsamic, hop, spice, wood

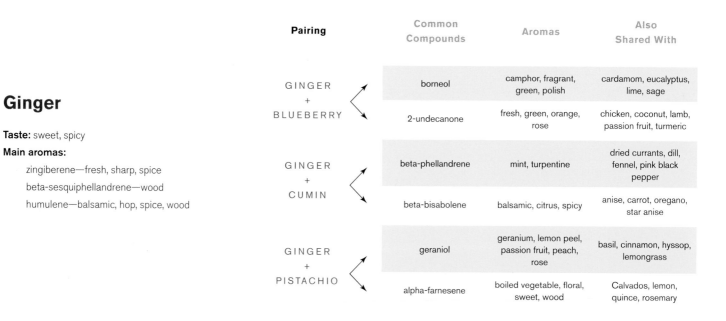

| Pairing | Common Compounds | Aromas | Also Shared With |
| --- | --- | --- | --- |
| GINGER + BLUEBERRY | borneol | camphor, fragrant, green, polish | cardamom, eucalyptus, lime, sage |
| | 2-undecanone | fresh, green, orange, rose | chicken, coconut, lamb, passion fruit, turmeric |
| GINGER + CUMIN | beta-phellandrene | mint, turpentine | dried currants, dill, fennel, pink black pepper |
| | beta-bisabolene | balsamic, citrus, spicy | anise, carrot, oregano, star anise |
| GINGER + PISTACHIO | geraniol | geranium, lemon peel, passion fruit, peach, rose | basil, cinnamon, hyssop, lemongrass |
| | alpha-farnesene | boiled vegetable, floral, sweet, wood | Calvados, lemon, quince, rosemary |

| Common Compounds | Aromas | Also Shared With | Pairing |
|---|---|---|---|
| 2-methylbutanal | almond, cocoa, fermented, hazelnut, malt | apple, beer, olives, rye bread | BARLEY + SWISS CHEESE |
| 4-hydroxypentanoic acid lactone | herb, sweet | beef, mushrooms, soybeans, tomato | |
| methyl dodecanoate | coconut, fat | alliums, cocoa, mustard, passion fruit | OAT + COCONUT |
| methylpyrazine | cocoa, green, hazelnut, popcorn, roasted | cashews, chicken, peanuts, sesame seeds, shrimp | |
| geranylacetone | green, hay, magnolia | cream, macadamias, sage, tomato | RICE + CLAM |
| 2-nonenal | paper | asparagus, ginger, olives, peas, pork | |

# Grain

**Taste:** bitter, slightly sweet

**Main aromas:**

isobutyraldehyde—caramel, cocoa, green, malt, nut

3-methylbutanal—almond, cocoa, fermented, hazelnut, malt

heptanal—citrus, dried fish, fat, green, nut

| Common Compounds | Aromas | Also Shared With | Pairing |
|---|---|---|---|
| ethyl tetradecanoate | ether, soap, wax | anise, apple, beer, coconut, passion fruit | GRAPE + PUMPKIN |
| ethyl acetate | brandy, contact glue, grape, sweet | alliums, beet, chestnuts, pumpkin, radish | |
| 3-methylphenol | feces, medicinal, phenol, sharp, urine | cheese, cocoa, hazelnuts, mushrooms, rum | GRAPE + PORK |
| acetophenone | almond, flower, meat, must, plastic | cauliflower, clams, pomegranate, vanilla | |
| 4-hydroxynonanoic acid, lactone | apricot, cocoa, coconut, peach, sweet | asparagus, beef, chile, corn, watercress | GRAPE + ELDERBERRY |
| linalool oxide D | citrus, floral, fragrant, green, sweet | fig, lemon balm, sage, strawberries | |

# Grape

**Taste:** sweet, tart

**Main aromas:**

1-hexanol—cooked vegetable, flower, green, herb, rotten

(Z)-3-hexen-1-ol—bell pepper, grass, green leaf, herb, unripe banana

beta-ionone—cedar, floral, raspberry, seaweed, violet

caprylaldehyde—citrus, fat, floral, green, pungent

## Green Bean

**Taste:** bitter

**Main aromas:**

n-hexanol—cooked vegetable, flower, green, herb, rot

n-hexanal—fresh, fruit, grass, green, oil

1-octen-3-ol—cucumber, earth, fat, floral, mushroom

| Pairing | Common Compounds | Aromas | Also Shared With |
|---|---|---|---|
| GREEN BEAN + VANILLA | vanillin | sweet, vanilla | cinnamon, fish, fennel, pineapple, pork, |
| | 1-phenylethanone | almond, flower, meat, must, plastic | alliums, corn, lentils, walnuts |
| GREEN BEAN + HAZELNUT | 2-pentylfuran | butter, floral, fruit, green bean | artichoke, jalapeño, parsley, shrimp |
| | 2-(sec-butyl)-3-methoxypyrazine | bell pepper, carrot, earth, green | beet, cucumber, parsnip, pumpkin |
| GREEN BEAN + DILL | beta-damascenone | cooked apple, floral, fruit, honey, tea | crème fraîche, lemon, savory, vinegar |
| | (Z)-3-hexen-1-ol | bell pepper, grass, green, herb, unripe banana | chervil, endive, peas, radish |

## Honey

**Taste:** sweet

**Main aromas:**

1-(2-furanyl)-ethanone—balsamic, cocoa, coffee, smoke, tobacco

2,3-butanediol—cream, floral, fruit, herb, onion, rubber

3-hydroxy-2-butanone—butter, cream, green bell pepper, rancid, sweat

| Pairing | Common Compounds | Aromas | Also Shared With |
|---|---|---|---|
| HONEY + SAGE | 2-methyl-1-propanol | apple, bitter, cocoa, fusel oil, plastic, solvent | apple, mezcal, tamarind, vinegar |
| | carvacrol | caraway, spice, thyme | lemon balm, nutmeg, orange, pistachios |
| HONEY + JALAPEÑO | methyl 2-methylbutanoate | apple, fruit, green apple, strawberry, sweet | cherries, lime, persimmon, prickly pear, thyme |
| | linalool | bergamot, cilantro, lavender, lemon, rose | apricot, cardamom, cilantro, juniper, mint, star anise |
| HONEY + MUSHROOM | 2-phenylethanol | fruit lilac, rose, wine | asparagus, barley, chervil, endive, mustard, watercress |
| | 4-methoxybenzaldehyde | almond, anise, mint, sweet | anise, basil, dates, fennel |

| Common Compounds | Aromas | Also Shared With | Pairing |
|---|---|---|---|
| diacetyl | butter, caramel, fruit, pastry, sweet | beet, cider, turkey, vinegar | JERUSALEM ARTICHOKE + CHERRY |
| hexadecane | root, earthy | eggplant, citrus, mustard, thyme | |
| beta-farnesene | citrus, oil, wood | carrot, cumin, lemon balm, pomegranate | JERUSALEM ARTICHOKE + GIN |
| pyridine | burnt, pungent, rancid | cabbage, mushrooms, shrimp, soy sauce | |
| propylbenzene | mothball | garlic, honey, peanuts, watercress | JERUSALEM ARTICHOKE + CRAB |
| anisole | anise | beef, cream, poblano, truffle | |

## Jerusalem Artichoke

**Taste:** sweet, slightly bitter

**Main aromas:**

beta-bisabolene—balsamic, wood

alpha-pinene—cedar, pine, resin, sharp, turpentine

terpinolene—sweet, pine, citrus

| Common Compounds | Aromas | Also Shared With | Pairing |
|---|---|---|---|
| ethyl butanoate | apple, butter, cheese, pineapple, strawberry | melon, orange, passion fruit, vinegar | KIWI + WHISKEY |
| styrene | balsamic, gasoline, plastic, rubber, solvent | vacciniums, cinnamon, peach, pecans | |
| 2-undecenal | sweet | caviar, chestnuts, lamb, yogurt | KIWI + CILANTRO |
| caprylaldehyde | citrus, fat, floral, green, pungent | lime, ginger, corn, sesame seeds | |
| 3-carene | bell pepper, lemon, resin, rubber | basil, cardamom, fennel, parsley, black pepper | KIWI + JALAPEÑO |
| (E)-2-hexen-1-ol | bell pepper, fat, geranium, green, spice | celery, dill, radish, tamarind | |

## Kiwi

**Taste:** sour, sweet

**Main aromas:**

ethyl butanoate—apple, butter, pineapple, red fruit, strawberry

(E)-2-hexenal—fat, floral, fruit, green grass, pungent

hexanal—fresh, fruit, grass, green, oil

# Lamb

**Taste:** umami, slightly sweet

**Main aromas:**

(E,E)-2,4-decadienal—cilantro, deep-fried, fat, oil, oxidized

(E)-2-nonenal—cucumber, cut grass, fat, paper, watermelon

4-methyl-3-thiazoline—burnt, earth, garlic

| Pairing | Common Compounds | Aromas | Also Shared With |
|---|---|---|---|
| LAMB + SHRIMP | 2,3,5-trimethylpyrazine | cocoa, earth, must, potato, roast | almonds, asparagus, okra, sesame |
| | 2-methylpyridine | ash, sweat | beans, oats, wild rice |
| LAMB + STRAWBERRY | butyric acid | butter, cheese, rancid, sour, sweat | bread, mushrooms, tarragon, wine |
| | heptanal | citrus, dry fish, fat, green, nut, rancid | artichoke, chervil, eggplant, garlic, olives |
| LAMB + MINT | 5-hydroxydodecanoic acid lactone | fruit, sweet | blue cheese, macadamias, sherry |
| | heptanoic acid | apricot, floral, rancid, sour, sweat | coffee, lemon, licorice, sauerkraut |

# Lemongrass

**Taste:** bland; flavor is fully dependent on aroma

**Main aromas:**

neral—citrus, lemon

geranial—flower, fruit, lemon, mint, pungent

beta-myrcene—balsamic, fruit, geranium, herb, must

| Pairing | Common Compounds | Aromas | Also Shared With |
|---|---|---|---|
| LEMONGRASS + CAULIFLOWER | limonene | balsamic, citrus, fragrant, fruit, greenery, herb | cherries, clams, kiwi, peach, pomegranate |
| | nonanal | citrus, fat, floral, green, paint | chicken, cilantro, mustard, strawberries |
| LEMONGRASS + CHERVIL | geraniol | geranium, lemon peel, passion fruit, peach, rose | alliums, basil, cheese, tomato |
| | beta-ciral | citrus, lemon | cardamom, curry, ginger, sage, thyme |
| LEMONGRASS + ROSEMARY | beta-myrcene | balsamic, fruit, geranium, herb, must | anise, carrot, dill, pistachios, star anise |
| | piperitone | fresh, mint | cinnamon, mint, black pepper, raspberries |

| Common Compounds | Aromas | Also Shared With | Pairing |
|---|---|---|---|
| tridecane | lime oil | dill, eggplant, lamb, rice, walnuts | LETTUCE + MUSTARD |
| phenethyl isothiocyanate | pungent | broccoli, horseradish, mushrooms, wasabi | |
| 2-methyl-1-propanol | apple, bitter, cocoa, fusel oil, plastic, solvent | alliums, apple, honey, mango, sake | LETTUCE + TAMARIND |
| 2-phenylethanol | fruit, honey, lilac, rose, wine | blue cheese, cashews, elderberries, peanuts, truffle | |
| 3-pentanol | fruit, green | chicken, oats, olives, peas, tomato | LETTUCE + SCALLOP |
| 1,2-dimethylbenzene | geranium | artichoke, beans, dill, persimmon, walnuts | |

## Lettuce

**Taste:** lightly bitter, sweet

**Main aromas:**

alpha-copaene—spice, wood

2-ethyl-1-hexanol—green, citrus, oil, rose

(E)-2-hexen-1-ol—bell pepper, fat, geranium, green, spice

| Common Compounds | Aromas | Also Shared With | Pairing |
|---|---|---|---|
| 1-penten-3-one | green, herb, metal, mustard, pungent | dates, endive, tea | MELON + CLAM |
| 1-hexanol | cooked vegetable, flower, grass, herb, rot | cilantro, hazelnuts, mustard, toasted bread | |
| isopentyl acetate | apple, banana, glue, pear | brandy, cocoa, sake | MELON + OLIVE |
| (E)-2-nonenal | cucumber, cut grass, fat, paper, watermelon | camomile, cucumber, fish, pecans | |
| nonanal | citrus, fat, floral, green, paint | broccoli, caraway, pistachios, pork | MELON + CORN |
| (E)-beta-ionone | cedar, floral, raspberry, seaweed, violet | citrus, fenugreek, seaweed, watercress | |

## Melon

**Taste:** sweet

**Main aromas:**

isobutyl acetate—apple, banana, floral, herb, plastic

(Z)-6-nonenal—green, cucumber, cantaloupe

(Z,Z)3,6-nonadienal—fat, soap, watermelon

isobutyl acetate—apple, banana, floral, herb, plastic

## Mollusk

**Taste:** salty, slightly sweet

**Main aromas:**

OYSTERS

1-octen-3-one—earth, metal, mushroom

(Z)-1,5-octadien-3-one—geranium, green, metal

MUSSELS/CLAMS/COCKLES

2-acetyl-2-thiazoline—caramel, popcorn, roast beef

diacetyl—butter, caramel

(Z)-4-heptenal—cream, fat, fish

OCTOPUS/SQUID

N,N-dimethylformamide—fish, pungent

acetoin—butter, cream, green bell pepper, rancid, sweat

SCALLOP

dimethyl sulfide—cabbage, gasoline, sulfur, wet earth

| Pairing | Common Compounds | Aromas | Also Shared With |
|---|---|---|---|
| CLAM + CAULIFLOWER | dimethylamine | fish | barley, celery, radish, wine |
| | 2-butanone | ether, fragrant, fruit, pleasant, sweet | beans, chile, garlic, thyme |
| MUSSELS + VANILLA | benzene | solvent | basil, ginger, parsley, pineapple |
| | methyl dodecanoate | coconut, fat | coconut, elderberries, mustard, wine |
| SQUID + ZUCCHINI | 1-Penten-3-ol | burnt, fish, green, meat, wet earth | cherries, ginger, lime, okra, tomato |
| | 2-methylbutanal | almond, cocoa, fermented, hazelnut, malt | mint, pistachios, pomegranate, vinegar |
| SCALLOP + ASPARAGUS | acetaldehyde | ether, floral, green apple, pungent, sweet | egg, lentils, peas, sake |
| | 2-nonanone | fragrant, fruit, green, hot milk, soap | bacon, caviar, lemongrass, rice |

## Mushroom

**Taste:** umami, slightly bitter, sweet

**Main aromas:**

3-octanol—citrus, moss, mushroom, nut, oil

benzaldehyde—bitter almond, burnt sugar, cherry, malt, roasted pepper

1,8-cineol—camphor, cool, eucalyptus, mint, sweet

| Pairing | Common Compounds | Aromas | Also Shared With |
|---|---|---|---|
| MUSHROOM + STRAWBERRY | 4-hydroxydodecanoic acid, gamma-lactone | apricot, flower, fruit, peach, sweet | cheese, chervil, chicken, pork, rum |
| | ethyl hexanoate | apple peel, brandy, fruit gum, overripe fruit, pineapple | apple, cloves, corn, tomato |
| MUSHROOM + HAZELNUT | 1-octen-3-one | earth, metal, mushroom | alliums, basil, mustard, sesame seeds |
| | 4-pentanolide | herb, sweet | barley, dates, soy sauce, wild rice |
| MUSHROOM + COCONUT | phenylacetaldehyde | berry, geranium, honey, nut, pungent | beet, celery, chile, tamarind, |
| | 2,3,5-trimethylpyrazine | cocoa, earth, must, potato, roast | almonds, clams, cilantro, toasted bread |

| Common Compounds | Aromas | Also Shared With | Pairing |
|---|---|---|---|
| (E,E)-2,4-decadienal | cilantro, deep-fried, fat, oil, oxidized | apricot, camomile, chicken, chile, tea, tomato | PEANUT + LIME |
| furfural | almond, baked potatoes, bread, burnt, spice | almonds, cinnamon, strawberries, licorice | |
| acetophenone | almond, flower, meat, must, plastic | clams, lentils, soy, tamarind, | WALNUT + SHRIMP |
| 2-heptanone | bell pepper, blue cheese, green, nut, spice | coconut, ginger, pear | |
| 2,6-dimethylpyrazine | cocoa, coffee, green, roast beef | barley, beef, cheese, malt, soybeans | ALMOND + ASPARAGUS |
| 2-formylpyrrole | musty, beefy, coffee | alliums, beer, lettuce, popcorn | |
| 4-vinylphenol | phenolic, medicinal, sweet | corn, vanilla, popcorn, raspberries | PECAN + MANGO |
| syringaldehyde | mild plastic, woody, vanilla, sweet | lime, pork, rum, | |

# Nut

**Taste:** both sweet and bitter, depending on variety

**Main aromas:**

ALMOND

benzaldehyde—bitter almond, burnt sugar, cherry, malt, roasted pepper

furaneol—burnt, caramel, cotton candy, honey, sweet

(E,Z)-2,4-decadienal—fat, fish, fried, green, paraffin oil

WALNUT

hexanal—fresh, fruit, grass, green, oil

pentanal—almond, bitter, malt, oil, pungent

PEANUT

2-isopropyl-3-methoxypyrazine—bell pepper, earth, green, hazelnut, pea

2-acetyl-1-pyrroline—popcorn

PECAN

2-propionyl-1-pyrroline—popcorn, roast

3-methylbutanal—almond, cocoa, cooked vegetable, malt, spice

HAZELNUT

linalool—bergamot, cilantro, lavender, lemon, rose

3-methyl-4-heptanone—fruit, hazelnut

PISTACHIO

alpha-pinene—cedar, pine, resin, sharp, turpentine

alpha-terpinolene—pine, plastic, sweet

SESAME SEEDS

2,5-dimethylpyrazine—burnt plastic, cocoa, medicinal, roast beef, toasted nut

2-ethyl-4-methyl-1H-Pyrrole—nut, sweet

A
B
C
D
E
F
G
H
J
K
L
M
N
O
P
R
S
T
V

# Olive

**Taste:** salty, fat, can range from slightly sweet to bitter depending on variety and processing

**Main aromas:**

beta-damascenone—cooked apple, floral, fruit, honey, tea

nonanal—citrus, fat, floral, green, paint

(E)-dec-2-enal—fat, fish, hay, paint, tallow

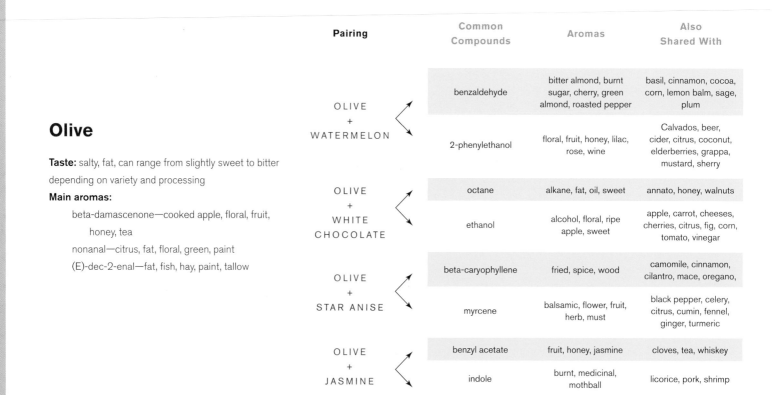

| Pairing | Common Compounds | Aromas | Also Shared With |
|---|---|---|---|
| OLIVE + WATERMELON | benzaldehyde | bitter almond, burnt sugar, cherry, green almond, roasted pepper | basil, cinnamon, cocoa, corn, lemon balm, sage, plum |
| | 2-phenylethanol | floral, fruit, honey, lilac, rose, wine | Calvados, beer, cider, citrus, coconut, elderberries, grappa, mustard, sherry |
| OLIVE + WHITE CHOCOLATE | octane | alkane, fat, oil, sweet | annato, honey, walnuts |
| | ethanol | alcohol, floral, ripe apple, sweet | apple, carrot, cheeses, cherries, citrus, fig, corn, tomato, vinegar |
| OLIVE + STAR ANISE | beta-caryophyllene | fried, spice, wood | camomile, cinnamon, cilantro, mace, oregano, |
| | myrcene | balsamic, flower, fruit, herb, must | black pepper, celery, citrus, cumin, fennel, ginger, turmeric |
| OLIVE + JASMINE | benzyl acetate | fruit, honey, jasmine | cloves, tea, whiskey |
| | indole | burnt, medicinal, mothball | licorice, pork, shrimp |

| Common Compounds | Aromas | Also Shared With | Pairing |
|---|---|---|---|
| acetaldehyde | ether, floral, green apple, pungent, sweet | lemon, cucumber, parsley, tomato | GARDEN PEA + STRAWBERRY |
| methyl isovalerate | apple, fruit, pineapple | melon, mint, mushrooms, sage | |
| diacetyl | butter, caramel, fruit, pastry, sweet | beet, cauliflower, hazelnuts, scallops, white wine | PEAS + CARROTS |
| alpha-terpineol | anise, fresh, mint, oil, sweet | anise, cilantro, basil, black pepper, savory | |
| cumene | solvent | celery, mushrooms, olives, parsley, pork | SUGAR SNAP + SHRIMP |
| 2-pentylfuran | butter, floral, fruit, green bean | capsicums, clams, rice tomato | |
| 2-undecanone | fresh, green, orange, rose | alliums, asparagus, ginger, turmeric | PEA + COCONUT |
| ethyl caprylate | apricot, brandy, fat, floral, pineapple | lime, mustard, pork, yogurt | |

# Pea

**Taste:** sweet

**Main aromas:**

(E)-2-heptenal—almond, fat, fruit, metal, soap

(E)-2-octenal—fat, fish oil, green, nut, plastic

3-isopropyl-2-methoxypyrazine—bell pepper, earth, green, hazelnut

3-sec-butyl-2-methoxypyrazine—bell pepper, carrot, earth, green

# Pome Fruit

**Taste:** sweet, tart

**Main aromas:**

APPLE

1-hexanol—cooked vegetable, flower, green, herb, rotten

(E)-2-hexenal—fat, floral, fruit, green grass, pungent

butyl acetate—apple, banana, glue, pungent, sweet

PEAR

ethyl (E,Z)-2,4-decadienoate—metal, pear

2-methylpropyl acetate—apple, banana, floral, herb, plastic

(E)-beta-damascenone—cooked apple, floral, fruit, honey

QUINCE

alpha-farnesene—boiled vegetable, floral, sweet, wood

ethyl octanoate—apricot, brandy, fat, floral, pineapple

ethyl hexanoate—apple peel, brandy, fruit gum, overripe fruit, pineapple

| Pairing | Common Compounds | Aromas | Also Shared With |
|---|---|---|---|
| POME FRUIT + BASIL | anethole | sweet, anise, licorice, medicinal | hyssop, cilantro, fennel, lemon balm, rhubarb |
| | geranyl acetate | lavender, rose, sweet | celery, citrus, curry, eucalyptus, lovage |
| APPLE + CRAB | 1-butanol | fruit, medicinal, solvent | cream, dill, endive, mezcal |
| | palmitic acid | rancid, wax | coconut, corn, plum, thyme, wasabi |
| QUINCE + HAZELNUT | 2-pentanone | burnt plastic, ether, fruit, kerosene, pungent | banana, cheese, coffee, rum, strawberries |
| | toulene | glue, paint, solvent | annatto, citrus, ginger, pumpkin, tea |
| PEAR + SAGE | hexyl acetate | apple, banana, grass, herb, pear | camomile, poultry, pomegranate, rye bread |
| | alpha-terpineol | anise, fresh, mint, oil, sweet | cumin, curry, lemongrass, nutmeg, parsnip |

| Common Compounds | Aromas | Also Shared With | Pairing |
|---|---|---|---|
| 6-methyl-5-hepten-2-one | citrus, mushroom, black pepper, rubber, strawberry | lemon balm, lemongrass, pistachios, plum | POMEGRANATE + DARK CHOCOLATE |
| isopentyl acetate | apple, banana, glue, pear | cider, grapes, sake, strawberries | |
| 2-methyl-3-buten-2-ol | earth, herb, oil, wood | dried currants, yogurt, sage, tamarind | POMEGRANATE + CHICKEN |
| 2-ethyl-1-hexanol | citrus, green, oil, rose | cherries, lychee, mint, tea | |
| gamma-terpinene | bitter, citrus, gasoline, resin, turpentine | cardamom, fennel, basil, pistachios | POMEGRANATE + CAMOMILE |
| ethyl benzoate | camomile, celery, fat, flower, fruit | kiwi, mango, rum, whiskey | |

# Pomegranate

**Taste:** sweet, astringent

**Main aromas:**

(Z)-3-Octenyl acetate—green, melon, pear, tropical

alpha-terpineol—anise, fresh, mint, oil, sweet

hexanol—cooked vegetable, flower, green, herb, rot

| Common Compounds | Aromas | Also Shared With | Pairing |
|---|---|---|---|
| 2-acetylfuran | balsamic, cocoa, coffee, smoke, tobacco | alliums, almonds, cocoa, sweet potato, oyster | PORK + APPLE |
| 1-heptanol | chemical, green, putrid, wood | citrus, mustard, olives, sherry | |
| 1H-pyrrole | nut, sweet | coffee, egg, mushrooms, tamarind | PORK + HAZELNUT |
| 2-furanmethanethiol | coffee, roasted meat | beef, peanuts, popcorn, sesame seeds | |
| 2-acetylthiazole | nut, popcorn, roast, sulfur | artichoke, beer, corn, cabbage | PORK + CLAM |
| 2,3,5-trimethylpyrazine | cocoa, earth, must, potato, roast | barley, coconut, okra, peanuts, whiskey | |
| 4,5-dimethylthiazole | green, nut, roasted, smoke | malt, oats, peach, scallops | PORK + COFFEE |
| 2-acetylthiophene | sulfur | asparagus, tomato, shrimp paste, bread | |

# Pork

**Taste:** umami, sweet

**Main aromas:**

2-pentylfuran—butter, floral, fruit, green bean

hexanal—fresh, fruit, grass, green, oil

1-octen-3-ol—cucumber, earth, fat, floral, mushroom

## Potato

**Taste:** bland; flavor is fully dependent on aroma

**Main aromas:**

(Z)-4-heptenal—biscuit, cream, fat, fish, rotten

2,3-diethyl-5-methylpyrazine—earth, meat, potato, roast

3,5-diethyl-2-methylpyrazine—baked, cocoa, roast, rum, sweet

| Pairing | Common Compounds | Aromas | Also Shared With |
|---|---|---|---|
| POTATO + CAVIAR | 2-propenal | burnt, sweet, pungent | beer, carrot, olives, wine |
| | 2,4-heptadienal | cucumber, fat, green, nut | corn, peas, soybeans, walnuts |
| POTATO + EGG | 2-methylbutanal | almond, cocoa, fermented, hazelnut, malt | alliums, apple, laurel, sake |
| | 2-decanone | fat, fruit | beef, cheese, mushrooms, shrimp |
| POTATO + AVOCADO | diacetyl | butter, caramel, fruit, pastry, sweet, yogurt | beet, blue cheese, pumpkin, scallops |
| | 2-pentylfuran | butter, floral, fruit, green bean | broccoli, chayote, pistachios, chicken |

## Poultry

**Taste:** umami, slightly sweet

**Main aromas:**

CHICKEN, TURKEY, GUINEA HEN

2-methyl-3-furanthiol—fried, meat, nut, potato, roasted meat

2-furfurylthiol—coffee, roasted meat

(E,Z)-2,4-decadienal—fat, fish, fried, green, paraffin oil

1-octen-3-one—earth, metal, mushroom

DUCK, PHEASANT, QUAIL, PIGEON (SQUAB), GOOSE

nonanal—citrus, fat, floral, green, paint

2-(E)-decenal—fat, fish, hay, paint, tallow

2-(E)-heptenal—almond, fat, fruit, metal, soap

| Pairing | Common Compounds | Aromas | Also Shared With |
|---|---|---|---|
| TURKEY + APRICOT | p-cymene | citrus, fresh, gasoline, solvent | anise, dried currants, basil, pistachios, turmeric |
| | benzyl alcohol | boiled cherry, moss, baked bread, rose, sweet | citrus, cinnamon, mustard, sage |
| DUCK + DILL | piperonal | anise, sweet | melon, vacciniums, black pepper, sherry, vanilla |
| | ethylbenzene | gasoline | cheese, egg, olives, tomato, walnuts |
| CHICKEN + BANANA | undecane | resin | chervil, cream, endive, garlic, walnuts |
| | (E)-2-hexenal (=leaf aldehyde) | fat, floral, fruit, green grass, pungent | arugula, leek, olives, radish |
| POULTRY + CRAWFISH | (E,E)-2,4-decadienal | cilantro, deep-fried, fat, oil, oxidized | chile, corn, mushrooms, pecans, sweet potato |
| | caprylaldehyde | citrus, fat, floral, green, pungent | artichoke, caraway, cilantro, ginger, oyster |

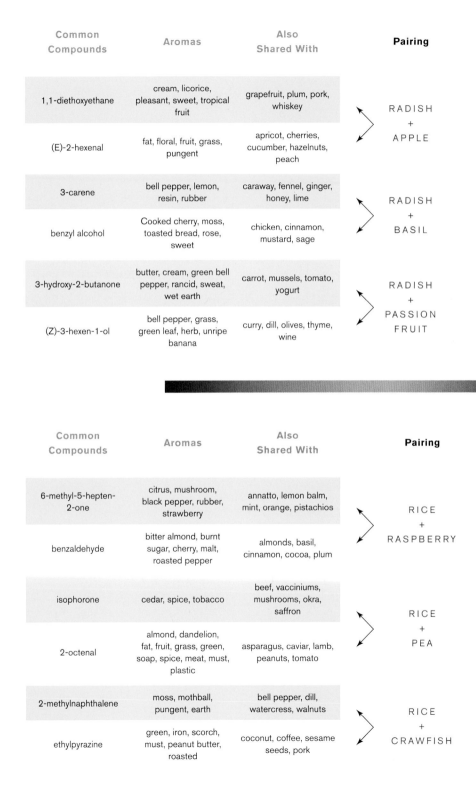

| Common Compounds | Aromas | Also Shared With | Pairing |
|---|---|---|---|
| 1,1-diethoxyethane | cream, licorice, pleasant, sweet, tropical fruit | grapefruit, plum, pork, whiskey | RADISH + APPLE |
| (E)-2-hexenal | fat, floral, fruit, grass, pungent | apricot, cherries, cucumber, hazelnuts, peach | |
| 3-carene | bell pepper, lemon, resin, rubber | caraway, fennel, ginger, honey, lime | RADISH + BASIL |
| benzyl alcohol | Cooked cherry, moss, toasted bread, rose, sweet | chicken, cinnamon, mustard, sage | |
| 3-hydroxy-2-butanone | butter, cream, green bell pepper, rancid, sweat, wet earth | carrot, mussels, tomato, yogurt | RADISH + PASSION FRUIT |
| (Z)-3-hexen-1-ol | bell pepper, grass, green leaf, herb, unripe banana | curry, dill, olives, thyme, wine | |

## Radish

**Taste:** bitter, spicy

**Main aromas:**

phytol—floral, balsamic

hexadecanoic acid—rancid, wax

4-(methylthio)butyl isothiocyanate—cabbage, radish

| Common Compounds | Aromas | Also Shared With | Pairing |
|---|---|---|---|
| 6-methyl-5-hepten-2-one | citrus, mushroom, black pepper, rubber, strawberry | annatto, lemon balm, mint, orange, pistachios | RICE + RASPBERRY |
| benzaldehyde | bitter almond, burnt sugar, cherry, malt, roasted pepper | almonds, basil, cinnamon, cocoa, plum | |
| isophorone | cedar, spice, tobacco | beef, vacciniums, mushrooms, okra, saffron | RICE + PEA |
| 2-octenal | almond, dandelion, fat, fruit, grass, green, soap, spice, meat, must, plastic | asparagus, caviar, lamb, peanuts, tomato | |
| 2-methylnaphthalene | moss, mothball, pungent, earth | bell pepper, dill, watercress, walnuts | RICE + CRAWFISH |
| ethylpyrazine | green, iron, scorch, must, peanut butter, roasted | coconut, coffee, sesame seeds, pork | |

## Rice

**Taste:** bland; flavor is fully dependent on aroma

**Main aromas:**

2-acetyl-1-pyrroline—popcorn

2-nonenal—paper

(E,E)-2,4-decadienal—cilantro, deep-fried, fat, oil, oxidized

# Root Vegetable

**Taste:** bitter, slightly sweet

**Main aromas:**

CELERY ROOT

limonene—balsamic, citrus, fragrant, fruit, herb

beta-pinene—pine, polish, resin, turpentine, wood

(E)-beta-ocimene—citrus, herb, mold, sweet, warm

PARSNIP

2-isobutyl-3-methoxypyrazine—earth, floral, green bell pepper, pea

myristicin—balsamic, carrot, nutmeg, spice

2-(sec-butyl)-3-methoxypyrazine—bell pepper, carrot, earth, green

SALSIFY

valencene—citrus, green, oil, wood

phenylacetaldehyde—pungent, green floral, sweet

1-hexanol—cooked vegetable, flower, green, herb, rotten

| Pairing | Common Compounds | Aromas | Also Shared With |
|---|---|---|---|
| SALSIFY + DRIED CURRANT | cis-carveol | caraway, cool | caraway, cardamom, dill, black pepper, rosemary |
| | beta-selinene | herb | basil, celery, ginger, olives, star anise |
| PARSNIP + CILANTRO | terpinolene | pine, plastic, sweet | anise, lime, ginger, black pepper, almonds |
| | myristicin | balsamic, carrot, nutmeg, spice, warm | caraway, lovage, fennel, star anise |
| CELERIAC + VANILLA | propylbenzene | mothball | beef, shellfish, pecans, watercress |
| | methyl laurate | coconut, fat | chile, cocoa, coconut, mustard, rum |

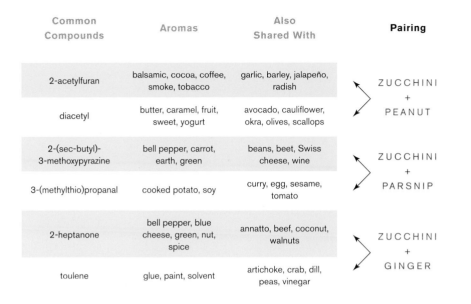

| Common Compounds | Aromas | Also Shared With | Pairing |
|---|---|---|---|
| 2-acetylfuran | balsamic, cocoa, coffee, smoke, tobacco | garlic, barley, jalapeño, radish | ZUCCHINI + PEANUT |
| diacetyl | butter, caramel, fruit, sweet, yogurt | avocado, cauliflower, okra, olives, scallops | |
| 2-(sec-butyl)-3-methoxypyrazine | bell pepper, carrot, earth, green | beans, beet, Swiss cheese, wine | ZUCCHINI + PARSNIP |
| 3-(methylthio)propanal | cooked potato, soy | curry, egg, sesame, tomato | |
| 2-heptanone | bell pepper, blue cheese, green, nut, spice | annatto, beef, coconut, walnuts | ZUCCHINI + GINGER |
| toulene | glue, paint, solvent | artichoke, crab, dill, peas, vinegar | |

## Squash, Summer

**Taste:** lightly bitter when raw, lightly sweet when cooked

**Main aromas:**

phenylacetaldehyde—berry, geranium, honey, nut, pungent

3-methylbutanal—almond, cocoa, cooked vegetable, malt, spice

leaf aldehyde—fat, floral, fruit, green grass, pungent

| Common Compounds | Aromas | Also Shared With | Pairing |
|---|---|---|---|
| 3-(methylthio)propanal | cooked potato, soy | apple, citrus, pistachios, scallops, sesame | PUMPKIN + MELON |
| 2-phenylethanol | fruit, honey, lilac, rose, wine | blue cheese, cinnamon, elderberries, mustard, tequila | |
| 2,3-pentanedione | bitter, butter, caramel, fruit, sweet | beer, cream, hazelnuts, tamarind | SWEET POTATO + CLAM |
| ethanol | almond, caramel, cooked, roasted garlic, spice | coconut, chile, coffee, peach | |
| 2-acetylfuran | balsamic, cocoa, coffee, smoke, tobacco | almonds, cherries, mushrooms, whiskey, peanuts | PUMPKIN + OKRA |
| 2,5-dimethylpyrazine | burnt plastic, cocoa, medicinal, roast beef, toasted nut | chestnuts, soy sauce, squid | |

## Squash, Winter

**Taste:** bitter when raw; sweet when cooked

**Main aromas:**

(Z)-3-hexenol—bell pepper, grass, green leaf, herb, unripe banana

n-Hexanol—cooked vegetable, flower, green, herb, rotten

dimethyl disulfide—cabbage, garlic, onion, putrid

limonene—balsamic, citrus, fragrant, fruit, herb

# Stone Fruit

**Taste:** sweet, sour

**Main aromas:**

### PEACH/NECTARINE

linalool—bergamot, cilantro, lavender, lemon, rose

benzaldehyde—bitter almond, burnt sugar, cherry, malt, roasted pepper

5-hydroxydecanoic acid lactone—coconut, fruit, peach, creamy

### PLUM, APRICOT

gamma-decalactone—apricot, fat, peach, pleasant, sweet

methyl octanoate—fruit, orange, sweet, wax, wine

hexyl acetate—apple, banana, confectionery, grass, herb

### CHERRY

benzaldehyde—bitter almond, burnt sugar, cherry, malt, roasted pepper

geranylacetone—green, hay, magnolia, fruit

benzyl alcohol—almond, cooked cherries, moss, baked bread, rose

| Pairing | Common Compounds | Aromas | Also Shared With |
|---|---|---|---|
| CHERRIES + BEER | isopentyl benzoate | balsamic, sweet | cocoa, lychee, quince, vinegar |
| | 2-hydroxybenzoic acid | sour, sparkling, sweet | cheddar cheese, honey, peanuts, vanilla |
| APRICOT + SAGE | alpha-phellandrene | citrus, mint, black pepper, turpentine, wood | cumin, fennel, lime, black pepper, pistachios |
| | linalool | bergamot, cilantro, floral, lavender, lemon, rose | chile, lemongrass, prickly pear, star anise |
| PLUM + ELDERBERRY | butyl hexanoate | fruit, grass, green | capers, melon, passion fruit, apple |
| | methyl 2-hydroxybenzoate | almond, caramel, medicinal, peppermint, sharp | basil, coffee, olives, strawberries |
| PEACH + SOY SAUCE | 4-hydroxypentanoic acid lactone | herb, sweet | chicken, hazelnuts, Swiss cheese, toasted bread |
| | ethyl benzoate | camomile, celery, fat, flower, fruit | mushrooms, sherry, tamarind, red wine |

| Common Compounds | Aromas | Also Shared With | Pairing |
|---|---|---|---|
| furaneol | burnt, caramel, cotton candy, honey, sweet | grapefruit, oats, peanuts, sherry | MOLASSES + MANGO |
| 2-methylbutanoic acid | butter, cheese, fermented, rancid, sour | chicken, coffee, mint, pistachios, rice | |
| 2-propanol | must, solvent | cider, lentils, pear, tofu, whiskey | MOLASSES + GARLIC |
| coumaran | sweet, balsamic | chestnuts, lemon balm, licorice, plum | |
| 2,6-dimethoxyphenol | medicinal, phenol, smoke | bonito flakes, pork, soy sauce | MOLASSES + FISH |
| propionic acid | fat, pungent, rancid, silage, soy | breadcrumb, Calvados, mushrooms, vinegar | |

## Sugar Syrup

**Taste:** sweet, bitter

**Main aromas (all):**

isopentyl hexanoate—anise, caramel, fruit, spice, yeast

ethyl phenylacetate—floral, fruit, honey, rose, sweet

isomaltol—burnt, caramel, fruit

MAPLE SYRUP

2-methyloxolan-3-one—sweet, rum, baked bread

vanillin—vanilla, sweet

MOLASSES

dimethyl sulfide—cabbage, gasoline, sulfur, wet earth

| Common Compounds | Aromas | Also Shared With | Pairing |
|---|---|---|---|
| 2-phenylethanol | fruit, honey, lilac, rose, wine | cheese, lemon, tamarind, truffle, watercress | TOMATO + COCONUT |
| hexanal | fresh, fruit, grass, green, oil | avocado, cardamom, cucumber, mussels, sesame | |
| methyl 2-hydroxybenzoate | almond, caramel, medicinal, peppermint, sharp | chile, honey, sage, vanilla | TOMATO + TEA |
| 4-methylpentanoic acid | floral | bread, capsicum, cheese, mushroom, mussels | |
| geranylacetone | green, hay, magnolia | clams, lemon balm, pork, rice | TOMATO + PASSION FRUIT |
| trans-linalool oxide | citrus, flower, fresh, lemon | cilantro, citrus, cocoa, strawberries | |
| dimethyl sulfide | cabbage, gasoline, sulfur, wet earth | crab, corn, okra, shrimp paste | TOMATO + CELERY |
| 4-methylacetophenone | bitter almond, floral, fruit, spice, sweet | mint, parsley, peach, black pepper | |

## Tomato

**Taste:** sweet, sour, umami

**Main aromas:**

3-methylbutanal—almond, cocoa, cooked vegetable, malt, spice

(Z)-3-hexenal—apple, bell pepper, cut grass, green, lettuce

3-methyl-1-butanol—banana, cocoa, floral, gasoline, malt

6-methyl-5-hepten-2-one—citrus, mushroom, black pepper, rubber, strawberry

# Tropical Fruit

**Taste:** sweet, sour

**Main aromas:**

MANGO

delta-3-carene—citrus, fir, pine needle

ethyl 3-methylbutanoate—apple, citrus, pineapple, sour, sweet

(Z)-3-hexenyl butanoate—apple, wine, green

PINEAPPLE

furaneol—burnt, caramel, cotton candy, honey, sweet, toasted

methyl 3-methylthiopropionate—vegetable, radish, horseradish

3-(methylthio)propanoic acid ethyl ester—sulfur, pineapple, onion, garlic, tomato

delta-octalactone—coconut, sweet, cream

BANANA

isoamyl acetate—apple, banana, glue, pear

isoamyl alcohol—banana, cocoa, floral, gasoline, malt

2-methyl-1-propanol—apple, bitter, cocoa, fusel oil, solvent

PASSION FRUIT

3-hydroxy-2-butanone—butter, cream, green bell pepper, rancid, sweat

2-heptanol—citrus, coconut, earth, fried, mushroom, oil

ethyl hexanoate—apple peel, brandy, fruit gum, overripe fruit, pineapple

COCONUT

delta-octalactone—coconut, peach

delta-nonalactone—nut, peach

ethyl octanoate—apricot, brandy, fat, floral, pineapple

PAPAYA

linalool—bergamot, cilantro, lavender, lemon, rose

isoamyl butanoate—green fruit

decanal—floral, fried, orange zest, penetrating, tallow

| Pairing | Common Compounds | Aromas | Also Shared With |
|---|---|---|---|
| BANANA + JALAPEÑO | isopentyl butanoate | green fruit | camomile, melon, pomegranate, whiskey |
| | isopentyl 3-methylbutanoate | fruit | beer, lettuce, mango, plum |
| COCONUT + LAMB | 5-hydroxyoctanoic acid lactone | coconut, peach | beef, chicken, sherry, tea, tomato |
| | 2-undecanone | fresh, green, orange, rose | corn, lemongrass, scallion, turmeric, |
| MANGO + DILL | (Z)-beta-ocimene | citrus, flower, herb, mold, warm spice | basil, cardamom, celery, fennel |
| | cis-3-Hexenyl acetate | banana, candy, floral, green | lemon, olives, peach, red wine |
| PINEAPPLE + BLUE CHEESE | butyric acid | butter, cheese, rancid, sour, sweat | mushrooms, red onion, rye bread |
| | methyl pentyl ketone | bell pepper, blue cheese, green, nut, spice | annatto, asparagus, pear, yogurt |
| PASSION FRUIT + MUSTARD | methyl benzoate | herb, lettuce, prune, sweet, violet | dried currants, kiwi, vanilla, vinegar |
| | citronellol | citrus, green, rose | cinnamon, ginger, thyme, rosemary |
| PAPAYA + TEQUILA | 2-methylthiophene | sulfur | crab, crawfish, popcorn, soybeans |
| | carvacrol | caraway, spice, thyme | cilantro, cumin, oregano, blueberries |

| Common Compounds | Aromas | Also Shared With | Pairing |
|---|---|---|---|
| ethanol | alcohol, floral, ripe apple, sweet | blue cheese, cocoa, pistachios, walnuts | TRUFFLE + BEET |
| dimethyl sulfide | cabbage, gasoline, sulfur, wet earth | asparagus, corn, peas, shellfish | |
| geranylacetone | green, hay, magnolia | brandy, hyssop, lemon, sage | TRUFFLE + TOMATO |
| 2-propenal | burnt sugar, pungent | beef, beer, potato, wine | |
| isobutanol | apple, bitter, cocoa, fusel oil, plastic, solvent | camomile, chicken, gin, radish | TRUFFLE + STRAWBERRY |
| 2-octanone | fat, fragrant, gasoline, mold, soap | chile, fig, cream, peanuts, vanilla | |

# Truffle

**Taste:** bland; flavor is fully dependent on aroma

**Main aromas:**

diacetyl—butter, caramel, fruit, sweet, yogurt

dimethyl disulphide—cabbage, garlic, onion, putrid

ethyl butyrate—apple, butter, cheese, sweet, pineapple, red fruit, strawberry

| Common Compounds | Aromas | Also Shared With | Pairing |
|---|---|---|---|
| vanillin | sweet, vanilla | citrus, dill, fish, sherry | VANILLA + CORN |
| 4-hydroxybenzaldehyde | almond, balsamic, vanilla | beer, pineapple, shrimp paste, tomato | |
| anisaldehyde | almond, anise, caramel, mint, popcorn, sweet | anise, basil, coffee, hazelnuts | VANILLA + TOMATO |
| 2,3-butanediol | cream, floral, fruit, herb, onion | cider, melon, pecans, vinegar | |
| 3-hydroxy-2-butanone | butter, cream, green bell pepper, rancid, sweat | barley, blue cheese, clams, honey | VANILLA + ASPARAGUS |
| acetovanillone | clove, flower, vanilla | beans, pork, soy sauce, wine | |

# Vanilla

**Taste:** slightly bitter

**Main aromas:**

vanillin—sweet, vanilla

p-hydroxybenzaldehyde—nut, almond, balsamic, vanilla

anisaldehyde—almond, anise, caramel, mint, sweet

# Additional Resources

## Books

BOUZARI, ALI. *INGREDIENT: UNVEILING THE ESSENTIAL ELEMENTS OF FOOD.* NEW YORK: HARPERCOLLINS/ ECCO, 2016.

CHARTIER, FRANÇOIS. *TASTE BUDS AND MOLECULES: THE ART AND SCIENCE OF FOOD, WINE, AND FLAVOR.* NEW YORK: HOUGHTON MIFFLIN HARCOURT, 2012.

*COGNITIVE COOKING WITH CHEF WATSON: RECIPES FOR INNOVATION FROM IBM & THE INSTITUTE OF CULINARY EDUCATION.* NAPERVILLE, IL: SOURCEBOOKS, 2015.

FORLEY, DIANE, WITH CATHERINE YOUNG. *THE ANATOMY OF A DISH.* NEW YORK: ARTISAN, 2002.

HARTINGS, MATTHEW. *CHEMISTRY IN YOUR KITCHEN.* CAMBRIDGE, UK: ROYAL SOCIETY OF CHEMISTRY, 2016.

MCGEE, HAROLD. *ON FOOD AND COOKING: THE SCIENCE AND LORE OF THE KITCHEN, COMPLETELY REVISED AND UPDATED.* NEW YORK: SCRIBNER, 2004.

PAGE, KAREN, AND ANDREW DORNENBURG. *THE FLAVOR BIBLE: THE ESSENTIAL GUIDE TO CULINARY CREATIVITY.* NEW YORK: LITTLE, BROWN AND COMPANY, 2008.

SHEPARD, GORDON M. *NEUROGASTRONOMY: HOW THE BRAIN CREATES FLAVOR AND WHY IT MATTERS.* NEW YORK: COLUMBIA UNIVERSITY PRESS, 2011.

STUCKEY, BARB. *TASTE WHAT YOU'RE MISSING: THE PASSIONATE EATER'S GUIDE TO WHY GOOD FOOD TASTES GOOD.* NEW YORK: SIMON & SCHUSTER/FREE PRESS, 2012.

## Websites

**FOODB.CA**
FOOD COMPONENT DATABASE, FROM THE METABOLOMICS INNOVATION CENTRE

**VCF-ONLINE.NL**
VOLATILE COMPOUNDS IN FOOD DATABASE, FROM TRISKELION (BY SUBSCRIPTION ONLY)

**IBMCHEFWATSON.COM**
INTERACTIVE FOOD PAIRINGS, FROM IBM WITH *BON APPÉTIT*

# Acknowledgments

As I've compiled this information and have written up the results, I have been fortunate to have an excellent collaborator and partner in crime: my wife and creative partner, Brooke Parkhurst. Brooke is the person responsible for keeping everything in order. That responsibility extends far beyond this book; from my wandering thoughts and technical jargon within these pages, to the lives of the three people (me and our two children) who depend on her for just about everything, Brooke keeps us on track. As I spent over a year buried in chemistry tables and scientific papers, Brooke was always there to keep me tethered to the real world. She has been my translator throughout this process, from fixing my (excessive) typos to rearranging my thoughts so that everyone—experienced chefs and basic home cooks alike—can understand and learn from them.

This book would never have been possible without the vision, dedication, and commitment of our agent, Joy Tutela, and our editor, Alexander Littlefield. They believed in this book and our ability to create it when no one else did. For that, we will be forever grateful. We also must acknowledge the support and incredible opportunities for professional growth received thanks to Rick Smilow and the Institute of Culinary Education.

We also thank the great scientific and technical minds of Matt Hartings and Ali Bouzari, whose expertise allowed an eager chef to sound a good bit smarter than he may actually be. Last, but certainly not least, we thank Jan Tulp, who brought the Flavor Matrixes to life. For his willing Skypes at odd hours and unwavering dedication to respond to each and every one of hundreds of emails obsessing over the smallest details to help us reach the perfect design; this book would be only half complete without you.

A
B
C
D
E
F
G
H
J
K
L
M
N
O
P
R
S
T
V

# Index

Note: Page references in *italics* indicate photographs.

A
B
C
D
E
F
**G**
**H**
**J**
**K**
L
M
N
O
P
R
S
T
V

A
B
C
D
E
F
G
H
J
K
**L**
**M**
N
O
P
R
S
T
V

A
B
C
D
E
F
G
H
J
K
L
M
**N**
**O**
**P**
R
S
T
V

A
B
C
D
E
F
G
H
J
K
L
M
N
O
P
R
S
**T**
V